Multivariate Statistical Methods

Multivariate Statistical Methods: A Primer offers an introduction to multivariate statistical methods in a rigorous yet intuitive way, without an excess of mathematical details. In this fifth edition, all chapters have been revised and updated, with clearer and more direct language than in previous editions, and with more up-to-date examples, exercises, and references, in areas as diverse as biology, environmental sciences, economics, social medicine, and politics.

Features

- A concise and accessible conceptual approach that requires minimal mathematical background.
- Suitable for a wide range of applied statisticians and professionals from the natural and social sciences.
- Presents all the key topics for a multivariate statistics course.
- The R code in the appendices has been updated, and there is a new appendix introducing programming basics for R.
- The data from examples and exercises are available on a companion website.

This book continues to be a great starting point for readers looking to become proficient in multivariate statistical methods, but who might not be deeply versed in the language of mathematics. In this edition, we provide readers with conceptual introductions to methods, practical suggestions, new references, and a more extensive collection of R functions and code that will help them to deepen their toolkit of multivariate statistical methods.

Multivariate Statistical Methods

A Primer

Fifth Edition

Bryan F. J. Manly, Jorge A. Navarro Alberto,
and Ken Gerow

CRC Press
Taylor & Francis Group
Boca Raton London New York

CRC Press is an imprint of the
Taylor & Francis Group, an **informa** business

A CHAPMAN & HALL BOOK

Cover image credit: Jorge A. Navarro Alberto

Fifth edition published 2025
by CRC Press
2385 NW Executive Center Drive, Suite 320, Boca Raton FL 33431

Second edition published 1994
Third edition published 2004
Fourth edition published 2016

by CRC Press
4 Park Square, Milton Park, Abingdon, Oxon, OX14 4RN

CRC Press is an imprint of Taylor & Francis Group, LLC

ISBN: 9781032592008 (hbk)
ISBN: 9781032591971 (pbk)
ISBN: 9781003453482 (ebk)

DOI: 10.1201/9781003453482

Typeset in Palatino
by Apex CoVantage, LLC

Access the companion website: https://jnavarroresearch.wordpress.com/
software/

Contents

Preface

It is gratifying to be the authors of the fifth edition of the book *Multivariate Statistical Methods: A Primer*, knowing that the previous editions have been a preferred text for students and professionals seeking an introduction to multivariate statistical methods. Possibly the success of the book resides in its modest purpose: to be an introductory book on multivariate statistical methods, particularly directed to nonmathematicians, without intending to be comprehensive. The intention has always been to keep the details to a minimum while still conveying a sense of what can be done. In other words, it is a book to "get you going" in a particular area of statistical methods.

We assume that readers have a working knowledge of elementary statistics, including tests of significance using the normal, t, chi-squared, and F distributions, as well as analysis of variance and linear regression. The material covered in a typical first university course in statistics should be quite adequate in this respect. Some facility with algebra is also required to follow the equations in certain parts of the text, and understanding the theory of multivariate methods to some extent does require the use of matrix algebra. However, the amount needed is not great if some details are accepted on faith. Matrix algebra is summarized in Chapter 2, and anyone who masters this chapter will have a reasonable competency in this area.

To some extent, the chapters can be read independently of each other. The first five are preliminary because they are mainly concerned with general aspects of multivariate data rather than with specific techniques. Chapter 1 introduces some examples to motivate the analyses covered in the book. Chapter 2 covers matrix algebra, Chapter 3 is about graphical methods of various types, Chapter 4 covers tests of significance, and Chapter 5 introduces the measurement of distances between objects based on variables measured on those objects. These chapters should be reviewed before Chapters 6 through 12, which cover the most important multivariate procedures in current use. The final Chapter 13 contains some general comments about the analysis of multivariate data.

The chapters in this fifth edition are the same as those in the previous four editions, and in making changes we have continually kept in mind the original intention of the book, which was that it should be as short as possible and to do no more than take readers to the stage where they can begin to use multivariate methods in an intelligent manner.

The most significant modifications made in this edition have been in the exposition of concepts and mathematical deductions, which are presented in a more colloquial way and appealing more to geometric interpretation. Some sections were written from scratch, for a clearer and more concise exposition. An exhaustive update of the recommended bibliography was also made in each chapter, with new up-to-date readings that we think have a level of complexity equal to or slightly higher than this primer.

To reach a broader audience, particularly those with computational skills, R code was introduced in the fourth edition, and complemented by appendices that showed how to do all relevant analyses in R. We have added an independent appendix introducing the use of the R package; codes for analyses are in the appendices for Chapters 2 through 12. Despite our preference for R, readers can continue to use the statistical package of their choice.

Other notable changes in this edition can be seen in the appearance of graphs, some of which contain more detailed pictorial elements that will allow readers a better comprehension. All these plots were created using R, purposedly keeping sophistication to a minimum, to make those plots reproducible using other statistical packages. In this edition there are also three new sections. The problem of confronting missing data is now addressed in Chapter 2, as a preamble for the application of one method for coping with missing data in principal components analysis, discussed in Chapter 6. The use of similarity matrices in multivariate analysis is now introduced in Chapter 5.

All the data used in examples and exercises are accessible through the supplemental resources page linked to this book. The data have been carefully chosen to cover a wide spectrum of interests, but with particular attention to natural and social scientists. In the previous editions there were no proposed exercises for Chapters 1–3 thus, exercise sections in these chapters are now included. Some of these new exercises are labelled with an **R**; they are particularly directed to readers getting started with R, as depicted in the main appendix and in the appendices for Chapters 1–3. Finally, new exercises were also included in the other chapters.

We wish to express our gratitude to all readers of previous editions of the book for their valuable feedback. Their comments and suggestions have been the main motivation to finally arriving at this fifth edition. This goal would not be possible to reach without the excellent editorial and promotional support of Chapman and Hall/CRC Press during all over the years. Thank you!

Bryan F. J. Manly
Dunedin, New Zealand

Jorge A. Navarro Alberto
Kanasín, México

Ken Gerow
Asheville, USA

The Authors

Bryan F. J. Manly, PhD, was born in London, UK on May 27, 1944, and he is practically retired from academic work. His areas of interest are in statistical ecology, environmental statistics, computer intensive statistics, and general applied statistics. He is the author of over 200 papers and seven books that have been both fundamental statistical research, and applications to several related disciplines. Bryan's academic career began in 1966 as a statistician and one of the first computer programmers, at the British multinational manufacturer Fisons, marking the start of a brilliant career as a researcher and statistical consultant in several countries around the world: University of Salford (UK), University of Papua New Guinea, University of Otago (New Zealand), Louisiana State University, University of Wyoming, and WEST, Inc. (USA). Among other distinctions, he is an Elected Fellow of the Royal Society of New Zealand, and he was awarded as Distinguished Statistical Ecologist in the International Ecology Congress, held in Manchester, 1994. Bryan is an excellent connoisseur of home brewing and homemade wine; everybody praises his good hand in making peerless wine!

Jorge A. Navarro Alberto, PhD, is a professor emeritus at the Autonomous University of Yucatán, México, where he specialized in ecological and environmental statistics research. Dr. Navarro Alberto earned his PhD degree in statistics at the University of Otago, New Zealand. His academic career spanned more than 36 years, teaching statistics for biologists, marine biologists, and natural resource managers in Mexico, and he was also a visiting professor at the University of Wyoming, with a vast experience in teaching multivariate analysis courses for life scientists. He is the coauthor of the last edition of the book *Randomization, Bootstrap and Monte Carlo Methods in Biology* and the co-editor of *Introduction to Ecological Sampling*, published by CRC Press. After retirement, Jorge is still active in the professional and academic arenas, working as a (more relaxed) part-time statistical consultant and as one of the associate editors of the international journal *Environmental and Ecological Statistics*. He also member of the Mexican representation at the International Statistical Literacy Project, Finland.

Ken Gerow, PhD, recently retired from the University of Wyoming, where, as a professor of statistics for over 30 years, he taught statistics to quantitative scientists from many disciplines. Dr. Gerow earned his PhD degree in Statistics at Cornell University. He is the author or a coauthor of over 90 research articles, books, and book chapters, in topics ranging from the molecular and cellular world to the visible world around us (plant, animal, and human systems). Ken considers himself to be a parasitic biologist because he only publishes with other people's data.

1

The Material of Multivariate Analysis

1.1 Examples of Multivariate Data

The statistical methods in elementary texts are mostly univariate methods, because they are only concerned with analyzing variation in a single response variable. On the other hand, the whole point of a multivariate analysis is to consider several related variables simultaneously, with each one being equally important, at least initially. The following examples illustrate the potential value of this more general approach.

1.1.1 Example 1.1: Storm Survival of Sparrows

After a severe storm on February 1, 1898, a number of moribund sparrows were taken to Hermon Bumpus' biological laboratory at Brown University, Rhode Island. Subsequently, about half of the birds died, and Bumpus saw this as an opportunity to see whether he could find any support for Charles Darwin's theory of natural selection. To this end, he made eight morphological measurements on each bird and also weighed the birds. The results for five of the measurements are shown in Table 1.1, for females only.

From those data, Bumpus (1898) concluded that "the birds which perished, perished not through accident, but because they were physically disqualified, and that the birds which survived, survived because they possessed certain physical characters." Specifically, he found that the survivors "are shorter and weigh less . . . have longer wing bones, longer legs, longer sternums and greater brain capacity" than the nonsurvivors. He also concluded that "the process of selective elimination is most severe with extremely variable individuals, no matter in which direction the variation may occur. It is quite as dangerous to be above a certain standard of organic excellence as it is to be conspicuously below the standard." Stabilizing selection occurred so that individuals with measurements close to the average survived better than individuals with measurements far from the average.

The development of multivariate statistical methods had hardly begun in 1898, when Bumpus was writing. The correlation coefficient as a measure of the relationship between two variables was devised by Francis Galton in

DOI: 10.1201/9781003453482-1

TABLE 1.1

Body Measurements of Female Sparrows

Bird	Survivorship	X_1	X_2	X_3	X_4	X_5
1	S	156	245	31.6	18.5	20.5
2	S	154	240	30.4	17.9	19.6
3	S	153	240	31.0	18.4	20.6
4	S	153	236	30.9	17.7	20.2
5	S	155	243	31.5	18.6	20.3
6	S	163	247	32.0	19.0	20.9
7	S	157	238	30.9	18.4	20.2
8	S	155	239	32.8	18.6	21.2
9	S	164	248	32.7	19.1	21.1
10	S	158	238	31.0	18.8	22.0
11	S	158	240	31.3	18.6	22.0
12	S	160	244	31.1	18.6	20.5
13	S	161	246	32.3	19.3	21.8
14	S	157	245	32.0	19.1	20.0
15	S	157	235	31.5	18.1	19.8
16	S	156	237	30.9	18.0	20.3
17	S	158	244	31.4	18.5	21.6
18	S	153	238	30.5	18.2	20.9
19	S	155	236	30.3	18.5	20.1
20	S	163	246	32.5	18.6	21.9
21	S	159	236	31.5	18.0	21.5
22	NS	155	240	31.4	18.0	20.7
23	NS	156	240	31.5	18.2	20.6
24	NS	160	242	32.6	18.8	21.7
25	NS	152	232	30.3	17.2	19.8
26	NS	160	250	31.7	18.8	22.5
27	NS	155	237	31.0	18.5	20.0
28	NS	157	245	32.2	19.5	21.4
29	NS	165	245	33.1	19.8	22.7
30	NS	153	231	30.1	17.3	19.8
31	NS	162	239	30.3	18.0	23.1
32	NS	162	243	31.6	18.8	21.3
33	NS	159	245	31.8	18.5	21.7
34	NS	159	247	30.9	18.1	19.0

(Continued)

TABLE 1.1   (Continued)

Bird	Survivorship	X_1	X_2	X_3	X_4	X_5
35	NS	155	243	30.9	18.5	21.3
36	NS	162	252	31.9	19.1	22.2
37	NS	152	230	30.4	17.3	18.6
38	NS	159	242	30.8	18.2	20.5
39	NS	155	238	31.2	17.9	19.3
40	NS	163	249	33.4	19.5	22.8
41	NS	163	242	31.0	18.1	20.7
42	NS	156	237	31.7	18.2	20.3
43	NS	159	238	31.5	18.4	20.3
44	NS	161	245	32.1	19.1	20.8
45	NS	155	235	30.7	17.7	19.6
46	NS	162	247	31.9	19.1	20.4
47	NS	153	237	30.6	18.6	20.4
48	NS	162	245	32.5	18.5	21.1
49	NS	164	248	32.3	18.8	20.9

Source: Data from Bumpus, H.C., Biological Lectures, Marine Biology Laboratory, Woods Hole, MA, 1898.

Note: X_1 = total length, X_2 = alar extent, X_3 = length of beak and head, X_4 = length of humerus, and X_5 = length of keel of sternum (all in millimeters). Original measurements were in inches and millimeters. Variable Survivorship indicates Birds (1–21) that survived, S, and Birds (22–49) that did not, NS.

1877 (Rao, 1983). However, it was another 56 years before Harold Hotelling described a practical method for carrying out a principal components analysis, which is one of the simplest multivariate analyses that can usefully be applied to Bumpus' data. Bumpus did not even calculate standard deviations. Nevertheless, his methods of analysis were sensible. Many authors have reanalyzed his data and, in general, have confirmed his conclusions.

Taking the data as an example for illustrating multivariate methods, several interesting questions arise, in particular:

1. How are the various measurements related? For example, does a large value for one of the variables tend to occur with large values for the other variables?

2. Are there statistically significant differences between survivors and casualties for the mean values of the variables?

3. Do the survivors and casualties show similar amounts of variation for the variables?

4. If the survivors and casualties do differ in terms of the distributions of the variables, then is it possible to construct some function of these variables that separates the two groups? It would then be convenient if large values of the function tended to occur with the survivors, as the function would then apparently be an index of the Darwinian fitness of the sparrows.

1.1.2 Example 1.2: Egyptian Skulls

The data in Table 1.2 are measurements made on male skulls from the area of Thebes in Egypt (Thomson and Randall-Maciver, 1905). There are samples of 30 skulls from each of the early predynastic period (circa 4000 BC), the late predynastic period (circa 3300 BC), the 12th and 13th Dynasties (circa 1850 BC), the Ptolemaic period (circa 200 BC), and the Roman period (circa AD 150). Four measurements are available on each skull, as illustrated in Figure 1.1.

For this example, some interesting questions are as follows:

1. How are the four measurements related?
2. Are there statistically significant differences between the sample means for the variables, and if so, do these differences reflect changes over time in the shape and size of skulls?
3. Are there significant differences between the sample standard deviations for the variables, and if so, do these differences reflect gradual changes over time in the amount of variation?
4. Is it possible to construct a function of the four variables that in some sense describes the changes over time?

These questions are, of course, rather similar in spirit to the ones suggested for Example 1.1.

It will be seen later that there are differences between the five samples that can be explained partly as time trends. It must be said, however, that the reasons for the apparent changes are unknown. Migration of other races into the region may well have been the most important factor.

1.1.3 Example 1.3: Distribution of a Butterfly

Data from a study of 16 colonies of the butterfly *Euphydryas editha* in California and Oregon (McKechnie et al., 1975) are in Table 1.3. Here, there are four environmental variables (altitude, annual precipitation, and the minimum and maximum annual temperatures) and six genetic variables (percentage frequencies for different *Pgi* genes as determined by the technique of electrophoresis). *Pgi* is a protein that functions as an enzyme inside a biological cell, involved in the breakdown of glucose. For the purposes of this example,

TABLE 1.2

Measurement on Male Egyptian Skulls

Skull	Early predynastic				Late predynastic				12th and 13th Dynasties				Ptolemaic period				Roman period			
	X_1	X_2	X_3	X_4	X_1	X_2	X_3	X_4	X_1	X_2	X_3	X_4	X_1	X_2	X_3	X_4	X_1	X_2	X_3	X_4
1	131	138	89	49	124	138	101	48	137	141	96	52	137	134	107	54	137	123	91	50
2	125	131	92	48	133	134	97	48	129	133	93	47	141	128	95	53	136	131	95	49
3	131	132	99	50	138	134	98	45	132	138	87	48	141	130	87	49	128	126	91	57
4	119	132	96	44	148	129	104	51	130	134	106	50	135	131	99	51	130	134	92	52
5	136	143	100	54	126	124	95	45	134	134	96	45	133	120	91	46	138	127	86	47
6	138	137	89	56	135	136	98	52	140	133	98	50	131	135	90	50	126	138	101	52
7	139	130	108	48	132	145	100	54	138	138	95	47	140	137	94	60	136	138	97	58
8	125	136	93	48	133	130	102	48	136	145	99	55	139	130	90	48	126	126	92	45
9	131	134	102	51	131	134	96	50	136	131	92	46	140	134	90	51	132	132	99	55
10	134	134	99	51	133	125	94	46	126	136	95	56	138	140	100	52	139	135	92	54
11	129	138	95	50	133	136	103	53	137	129	100	53	132	133	90	53	143	120	95	51
12	134	121	95	53	131	139	98	51	137	139	97	50	134	134	97	54	141	136	101	54
13	126	129	109	51	131	136	99	56	136	126	101	50	135	135	99	50	135	135	95	56
14	132	136	100	50	138	134	98	49	137	133	90	49	133	136	95	52	137	134	93	53
15	141	140	100	51	130	136	104	53	129	142	104	47	136	130	99	55	142	135	96	52
16	131	134	97	54	131	128	98	45	135	138	102	55	134	137	93	52	139	134	95	47
17	135	137	103	50	138	129	107	53	129	135	92	50	131	141	99	55	138	125	99	51
18	132	133	93	53	123	131	101	51	134	125	90	60	129	135	95	47	137	135	96	54

(Continued)

TABLE 1.2 (Continued)

Skull	Early predynastic				Late predynastic				12th and 13th Dynasties				Ptolemaic period				Roman period			
	X_1	X_2	X_3	X_4	X_1	X_2	X_3	X_4	X_1	X_2	X_3	X_4	X_1	X_2	X_3	X_4	X_1	X_2	X_3	X_4
19	139	136	96	50	130	129	105	47	138	134	96	51	136	128	93	54	133	125	92	50
20	132	131	101	49	134	130	93	54	136	135	94	53	131	125	88	48	145	129	89	47
21	126	133	102	51	137	136	106	49	132	130	91	52	139	130	94	53	138	136	92	46
22	135	135	103	47	126	131	100	48	133	131	100	50	144	124	86	50	131	129	97	44
23	134	124	93	53	135	136	97	52	138	137	94	51	141	131	97	53	143	126	88	54
24	128	134	103	50	129	126	91	50	130	127	99	45	130	131	98	53	134	124	91	55
25	130	130	104	49	134	139	101	49	136	133	91	49	133	128	92	51	132	127	97	52
26	138	135	100	55	131	134	90	53	134	123	95	52	138	126	97	54	137	125	85	57
27	128	132	93	53	132	130	104	50	136	137	101	54	131	142	95	53	129	128	81	52
28	127	129	106	48	130	132	93	52	133	131	96	49	136	138	94	55	140	135	103	48
29	131	136	114	54	135	132	98	54	138	133	100	55	132	136	92	52	147	129	87	48
30	124	138	101	46	130	128	101	51	138	133	91	46	135	130	100	51	136	133	97	51

Source: Data from Thomson, A. and Randall-Maciver, P., *Ancient Races of the Thebaid*, Oxford University Press, Oxford, London, 1905. X_1 = maximum breadth, X_2 = basibregmatic height, X_3 = basialveolar length, X_4 = nasal height, all in millimeters.

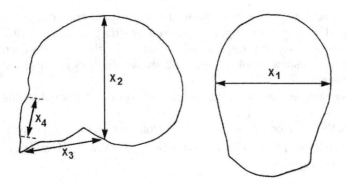

FIGURE 1.1

Four measurements made on Egyptian male skulls.

TABLE 1.3

Environmental Variables and Phosphoglucose-Isomerase (*Pgi*) Gene Frequencies for Colonies of the Butterfly *Euphydryas editha* in California and Oregon, USA

Colony	Altitude (feet)	Annual precipitation (inches)	Temperature (°F) Maximum	Minimum	Frequencies of *Pgi* mobility genes (%)[a] 0.4	0.6	0.8	1	1.16	1.3
SS	500	43	98	17	0	3	22	57	17	1
SB	808	20	92	32	0	16	20	38	13	13
WSB	570	28	98	26	0	6	28	46	17	3
JRC	550	28	98	26	0	4	19	47	27	3
JRH	550	28	98	26	0	1	8	50	35	6
SJ	380	15	99	28	0	2	19	44	32	3
CR	930	21	99	28	0	0	15	50	27	8
UO	650	10	101	27	10	21	40	25	4	0
LO	600	10	101	27	14	26	32	28	0	0
DP	1,500	19	99	23	0	1	6	80	12	1
PZ	1,750	22	101	27	1	4	34	33	22	6
MC	2,000	58	100	18	0	7	14	66	13	0
IF	2,500	34	102	16	0	9	15	47	21	8
AF	2,000	21	105	20	3	7	17	32	27	14
GH	7,850	42	84	5	0	5	7	84	4	0
GL	10,500	50	81	−12	0	3	1	92	4	0

Source: Data from McKechnie, S.W. et al., *Genetics*, 81, 571–94, 1975, with the environmental variables rounded to integers for simplicity.

Note: The original data were for 21 colonies, but for the present example, five colonies with small samples for the estimation of gene frequencies have been excluded to make all estimates about equally reliable.

[a] The numbers 0.40, 0.60, etc. represent different genetic types of *Pgi* so that the frequencies for a colony (adding to 100%) show the frequencies of the different types for the *E. editha* at that location.

there is no need to go into the details of how the gene frequencies were determined, and strictly speaking, they are not exactly gene frequencies anyway. It is sufficient to say that the frequencies describe the genetic distribution of the butterfly to some extent. Figure 1.2 shows the geographical locations of the colonies.

In this example, questions that can be asked include the following:

1. Are the *Pgi* frequencies similar for colonies that are close in space?
2. To what extent, if any, are the *Pgi* frequencies related to the environmental variables?

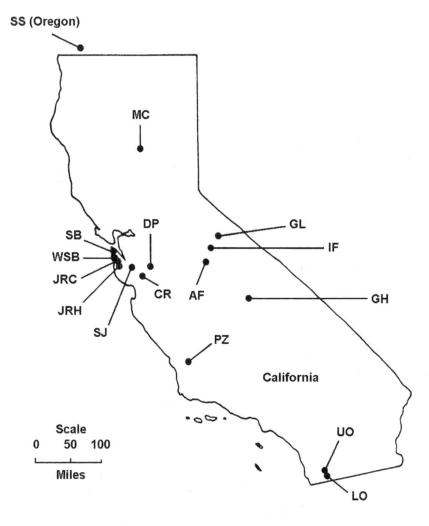

FIGURE 1.2
Colonies of *Euphydras editha* in California and Oregon.

If the genetic composition of the colonies was largely determined by past and present migration, then gene frequencies will tend to be similar for colonies that are close in space but may show little relationship with the environmental variables. On the other hand, if the environment is most important, this should show up in relationships between the gene frequencies and the environmental variables (assuming that the right variables have been measured), but neighboring colonies will only have similar gene frequencies if they have similar environments. Colonies that are close in space may have similar environments, so it may be difficult to reach a clear conclusion on this matter.

1.1.4 Example 1.4: Prehistoric Dogs from Thailand

Excavations of prehistoric sites in northeast Thailand have produced a collection of canid (dog) bones covering a period from about 3500 BC to the present. The prehistoric dog may have descended from the golden jackal (*Canis aureus*) or from the wolf, but wolves are not native to Thailand. The nearest indigenous sources are western China (*Canis lupus chanco*) and the Indian subcontinent (*Canis lupus pallides*).

To try to clarify the ancestors of the prehistoric dogs, mandible (lower jaw) measurements were made on the available specimens. These were then compared with the same measurements made on the golden jackal, the Chinese wolf, and the Indian wolf. The comparisons were also extended to include the dingo, which may have its origins in India; the cuon (*Cuon alpinus*), which is indigenous to Southeast Asia, and modern village dogs from Thailand (Highman et al., 1980).

Table 1.4 gives mean values for the six mandible measurements for specimens from all seven groups. What do the measurements suggest about the

TABLE 1.4

Mean Mandible Measurements for Seven Canine Groups

	X_1	X_2	X_3	X_4	X_5	X_6
Modern dog	9.7	21.0	19.4	7.7	32.0	36.5
Golden jackal	8.1	16.7	18.3	7.0	30.3	32.9
Chinese wolf	13.5	27.3	26.8	10.6	41.9	48.1
Indian Wolf	11.5	24.3	24.5	9.3	40.0	44.6
Cuon	10.7	23.5	21.4	8.5	28.8	37.6
Dingo	9.6	22.6	21.1	8.3	34.4	43.1
Prehistoric dog	10.3	22.1	19.1	8.1	32.2	35.0

Source: Data from Higham, C.F.W. et al., *J. Archaeol. Sci.*, 7, 149–65, 1980.
Note: X_1 = breadth of mandible, X_2 = height of mandible below the first molar, X_3 = length of the first molar, X_4 = breadth of the first molar, X_5 = length from first to third molar inclusive, X_6 = length from first to fourth premolar inclusive (all in millimeters).

relationships between the groups and, in particular, how do the prehistoric dog relate to the other groups.

1.1.5 Example 1.5: Cost of a Healthy Diet in Europe

In 2020, the Food and Agriculture Organization of the United Nations (FAO, 2023) began reporting global, regional, and country level indicators on the cost and affordability of a healthy diet (CoAHD). These indicators lie within "the framework of the so-called Sustainable Development Goals (SDGs) to end hunger, achieve food security and improved nutrition, and promote sustainable agriculture by 2030." A healthy diet provides adequate calories and adequate levels of all essential nutrients and food groups needed for an active and healthy life. CoAHDs have been estimated on a yearly basis in 143 countries, even for some years before 2020. They are measures of population's physical and economic access to least expensive locally available foods to meet requirements for a healthy diet (Herforth *et al.*, 2020, 2022).

Table 1.5 gives the cost of six food groups for 38 European countries in 2017. In this case, multivariate methods may be useful in isolating groups of countries with similar patterns of cost of a healthy diet and in comprehending

TABLE 1.5

Costs of Food Types Constituting a Healthy Diet in 38 Countries of Europe in 2017

Country*	Group	CSTS	CASF	CLNS	CVEG	CFRT	COFT
Albania	NEU	0.599	1.204	0.441	0.707	0.911	0.089
Austria	EU95	0.229	0.612	0.36	0.698	0.819	0.055
Belarus	NEU	0.506	0.826	0.643	0.349	0.678	0.174
Belgium	EU95	0.222	0.745	0.369	0.677	0.790	0.060
Bosnia and Herzegovina	NEU	0.540	0.921	0.512	0.825	0.916	0.132
Bulgaria	EU17	0.504	0.963	0.478	0.774	0.959	0.101
Croatia	EU17	0.496	0.955	0.556	1.138	0.950	0.073
Czechia	EU17	0.297	0.700	0.377	0.757	0.693	0.076
Denmark	EU95	0.235	0.498	0.263	0.78	0.546	0.053
Estonia	EU17	0.333	0.752	0.671	0.542	0.742	0.085
Finland	EU95	0.195	0.543	0.407	0.727	0.597	0.077
France	EU95	0.246	0.640	0.277	0.848	0.859	0.066
Germany	EU95	0.235	0.666	0.301	0.816	0.714	0.052
Greece	EU95	0.410	0.797	0.410	0.742	0.581	0.096
Hungary	EU17	0.407	0.805	0.420	0.754	0.828	0.089
Iceland	EEA	0.277	0.596	0.277	0.525	0.484	0.055

(Continued)

TABLE 1.5 (Continued)

Country*	Group	CSTS	CASF	CLNS	CVEG	CFRT	COFT
Ireland	EU95	0.326	0.523	0.285	0.582	0.621	0.060
Italy	EU95	0.221	0.778	0.266	0.751	0.812	0.057
Latvia	EU17	0.335	0.794	0.314	0.921	0.648	0.111
Lithuania	EU17	0.368	0.702	0.343	0.834	0.670	0.086
Luxembourg	EU95	0.208	0.656	0.280	0.530	0.762	0.056
Malta	EU17	0.508	0.795	0.450	0.730	0.931	0.080
Montenegro	NEU	0.425	0.862	0.367	0.829	0.833	0.081
Netherlands	EU95	0.322	0.614	0.255	0.892	0.607	0.053
North Macedonia	NEU	0.525	0.949	0.347	0.712	0.640	0.146
Norway	EEA	0.565	0.691	0.257	1.122	0.615	0.075
Poland	EU17	0.318	0.584	0.377	0.890	0.651	0.090
Portugal	EU95	0.271	0.583	0.275	0.620	0.687	0.078
Republic of Moldova	NEU	0.462	0.746	0.360	0.339	0.432	0.121
Romania	EU17	0.358	0.945	0.351	0.497	0.697	0.073
Russian Federation	NEU	0.360	0.980	0.191	0.739	0.745	0.135
Serbia	NEU	0.572	0.998	0.612	0.920	0.872	0.096
Slovakia	EU17	0.350	0.667	0.311	0.974	0.634	0.077
Slovenia	EU17	0.290	0.698	0.346	0.722	0.669	0.072
Spain	EU95	0.278	0.623	0.211	0.739	0.796	0.053
Sweden	EU95	0.470	0.644	0.347	0.766	0.773	0.085
Switzerland	NEU	0.212	0.675	0.249	0.743	0.577	0.067
United Kingdom	EU95	0.147	0.494	0.195	0.399	0.533	0.053

Source: Data from FAO (2023). * There were no data available for Ukraine.
Note: The six numeric variables contain the costs (measured as the purchasing power parity dollar per person per day or PPP dollar per person per day) of individual food groups that make up the Healthy Diet Basket: CSTS, starchy staples; CASF, animal source foods; CLNS, legumes, nuts, and seeds; CVEG; vegetables; CFRT, fruits; COFT, oils and fats. The country groups are: EU95, countries that joined the European Union (EU) until 1995; EU17, countries that joined the EU between 1996 and 2017; EEA, non-EU countries, members of the European Economic Area; NEU, "genuine" non-EU countries.

relationships between countries. Differences between countries can be analyzed depending on whether they belong to the European Union (EU) or not. From the former group a subdivision of countries was defined: the first countries that joined the EU until 1995 (EU95) and countries that made up the largest enlargement of the EU, between 1996 and 2017 (EU17). Although the

United Kingdom ceased to be a member state of the EU on January 31, 2020, it belonged to the EU when the CoAHD data were recorded. Two groups of non-EU countries are also distinguished: the European Economic Area (EEA) that allows Iceland and Norway to be part of the EU's single market, and "genuine" non-European Union countries (NEU).

1.2 Preview of Multivariate Methods

These examples are typical of the raw material for multivariate statistical methods. In all cases, there are several variables of interest, and these are clearly not independent of each other. At this point, we give a brief preview of how these examples will be used in the rest of this book.

Principal components analysis is designed to reduce the number of variables that need to be considered to a small number of indices (called the principal components) that are linear combinations of the original variables. For example, much of the variation in the body measurements of sparrows ($X_1 - X_5$) shown in Table 1.1 will be related to the general size of the birds, and the total

$$I_1 = X_1 + X_2 + X_3 + X_4 + X_5$$

should measure this aspect of the data quite well, accounting for one dimension of the data. Another index is

$$I_2 = X_1 + X_2 + X_3 - X_4 - X_5,$$

which contrasts the first three measurements with the last two. This reflects another dimension of the data. Principal components analysis provides an objective way of finding indices of this type so that the variation in the data can be accounted for as concisely as possible. Often, two or three principal components provide a good summary of all the original variables. Consideration of the values of the principal components may then make it much easier to understand what the data have to say. In short, principal components analysis is a means of simplifying data by reducing the number of variables.

Factor analysis also attempts to account for the variation in a number of original variables using a smaller number of index variables or factors. It is assumed that each original variable can be expressed as a linear combination of these factors, plus a residual term that reflects the extent to which the variable is independent of the other variables. For example, a two-factor model for the sparrow data assumes that

$$X_1 = a_{11}F_1 + a_{12}F_2 + e_1$$

$$X_2 = a_{21}F_1 + a_{22}F_2 + e_2$$

$$X_3 = a_{31}F_1 + a_{32}F_2 + e_3$$

$$X_4 = a_{41}F_1 + a_{42}F_2 + e_4$$

and

$$X_5 = a_{51}F_1 + a_{52}F_2 + e_5$$

where:

a_{ij} values are constants;

F_1 and F_2 are the factors;

e_i represents the variation in X_i that is independent of the variation in the other X variables.

Here, F_1 might be the factor of size. In that case, the coefficients a_{11}, a_{21}, a_{31}, a_{41}, and a_{51} would all be positive, reflecting the fact that some birds tend to be large and some birds tend to be small on all body measurements. The second factor F_2 might then measure an aspect of the shape of birds, with some positive coefficients and some negative coefficients. If this two-factor model fitted the data well, then it provides a relatively straightforward description of the relationship between the five body measurements.

One type of factor analysis starts by taking the first few principal components as the factors in the data being considered. These initial factors are then modified by a special transformation process called *factor rotation* to make them easier to interpret. Other methods for finding initial factors are also used. A rotation to simpler factors is almost always done.

Discriminant function analysis asks whether it is possible to separate different groups on the basis of the available measurements. This could be used, for example, to see how well surviving and nonsurviving sparrows can be separated using their body measurements (Example 1.1) or how skulls from different epochs can be separated, again using size measurements (Example 1.2). Like principal components analysis, discriminant function analysis is based on finding suitable linear combinations of the original variables to achieve the intended aim.

Cluster analysis tries to identify groups of similar objects. There is not much point in doing this with the data of Examples 1.1 and 1.2, as the groups (survivors/nonsurvivors and epochs) are already known. However, in Example 1.3, there might be some interest in grouping colonies on the basis of environmental variables or *Pgi* frequencies, while in Example 1.4, the main point of interest is in the similarity between prehistoric Thai dogs and other animals.

Likewise, in Example 1.5, the European countries might be grouped in terms of their similarity in patterns of costs of healthy food.

With *canonical correlation*, the variables (not the objects) are divided into two groups, and interest centers on the relationship between these. Thus, in Example 1.3, we can explore the relationship between, on one hand, the four environmental variables versus the six genetic variables. Finding relationships, if any, between these two groups of variables is of considerable biological interest.

Multidimensional scaling begins with data on some measure of the distances between a number of objects. From these distances, a map is then constructed showing how the objects are related. This can be used to measure how far apart pairs of objects are without having any idea of how the objects are related in a geometric sense. Thus, in Example 1.4, there are ways of measuring the distances between pairs of animal groups. There are 21 distances altogether, and from these distances, multidimensional scaling can be used to produce a type of map of the relationships between the groups. With a one-dimensional map, the groups are placed along a straight line. With a two-dimensional map, they are represented by points on a plane. With a three-dimensional map, they are represented by points within a cube. Four-dimensional and higher solutions are also possible, although these have limited use because they cannot be visualized in a simple way. The value of a one-, two-, or three-dimensional map is clear for Example 1.4, as such a map would immediately show to which groups prehistoric dogs are most similar. Multidimensional scaling may be a useful alternative to cluster analysis in this case. A map of European countries based on patterns of costs of healthy food might also be of interest in Example 1.5.

Principal components analysis and multidimensional scaling are sometimes referred to as methods for *ordination*. They are methods for producing axes against which a set of objects of interest can be plotted. Other methods of ordination are also available.

Principal coordinates analysis is like a type of principal components analysis that starts off with information on the extent to which the pairs of objects are different in a set of objects, instead of the values for measurements on the objects. As such, it is similar to multidimensional scaling. However, the assumptions made and the numerical methods used are not the same.

Correspondence analysis starts with data on the abundance of each of several characteristics for each of a set of objects. This is useful in ecology, for example, where the objects of interest are often different sites, the characteristics are different species, and the data consist of abundances of the species in samples taken from the sites. The purpose of correspondence analysis would then be to clarify the relationships between the sites, as expressed by species distributions, and the relationships between the species, as expressed by site distributions.

1.3 The Multivariate Normal Distribution

The normal distribution for a single variable should be familiar to readers of this book. It has the well-known bell-shaped frequency curve, and many standard univariate statistical methods are based on the assumption that data are normally distributed (or, at least, that the sample size is large enough that the central limit theorem [CLT] assures that the distribution of the mean is approximately normal).

Given the prominence of the normal distribution in univariate statistical methods, it will come as no surprise to discover that the multivariate normal distribution has a central position with multivariate statistical methods. Many of these methods require the assumption that the data being analyzed have multivariate normal distributions; note that this is a stricter assumption than that for univariate methods, wherein the CLT can come to the rescue.

The exact definition of a multivariate normal distribution is not too important. The approach of most people, for better or worse, seems to be to regard data as being normally distributed unless there is some reason to believe that this is not true. In particular, if all the individual variables being studied appear to be normally distributed, then it is assumed that the joint distribution is multivariate normal. The definition of multivariate normality requires more than this. Figure 1.3 shows a bivariate (joint) normal distribution: it looks like a two-cornered cocked hat with infinite brim, where slices

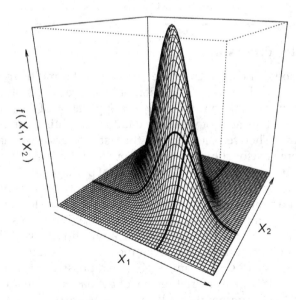

FIGURE 1.3
A bivariate normal distribution $f(X_1, X_2)$. Thicker curves highlight marginal (univariate) normal distributions.

perpendicular to the brim produce the well-known bell-shaped univariate (marginal) normal distributions, while the top of the crown is positioned on the mean values of X_1 and X_2. The shape of the cocked hat is determined by the linear correlation between X_1 and X_2, covered in Chapter 2: the smaller the linear correlation between X_1 and X_2, the rounder the crown of the cocked hat will be. Figure 1.3 exemplifies an important aspect of a multivariate normal distribution: that it is specified completely by a mean vector and a covariance matrix. The definitions of these are given in Section 2.7. Basically, the mean vector contains the mean values for all the variables being considered, while the covariance matrix contains the variances for all the variables plus the covariances, which measure the extent to which all pairs of variables are related. This last feature of the data often suggests that the use of multivariate statistical methods will better than univariate statistics, as the potential relations between variables are incorporated in the analyses.

Cases arise where the assumption of multivariate normality is clearly invalid. For example, one or more of the variables being studied may have a highly skewed distribution with several very high (or low) values, there may be many repeated values, and so on. Another situation on non-normal distributions arises when you have a very small sample, and the variables of interest are discrete. These problems can sometimes be overcome by an appropriate transformation of the data, as discussed in elementary texts on statistics. If this does not work, then a rather special form of analysis may be required.

1.4 Computer Programs

Practical methods for carrying out the calculations for multivariate analyses have been developed since the inception of bivariate statistical methods by Francis Galton in the 19th century. However, application of these methods for more than a small number of variables had to wait until computers became readily available. Therefore, it is only in the last 50 years or so that the methods have become reasonably easy to carry out for the average researcher.

Nowadays, there are many standard statistical packages and computer programs available for calculations on computers of all types. It is intended that this book should provide readers with enough information to use any of these packages and programs intelligently. Easier to use (it requires no programming), Minitab software (Minitab, Inc., 2023) will do many of the analyses we consider in this text. However, given the wide accessibility of the R programming language (R Core Team, 2024), we emphasize its use here in more detail. The Appendix gives a brief review of the basics of the R environment needed to run the R commands in all chapters. More extensive treatments of the R language can be found in many books and manuals, including those by Ekstrøm (2016), Davies (2016), Long and Teetor (2019), Kabacoff (2020), Zamora

Saiz et al. (2020), Jones et al. (2022), and Venables et al. (2024). The book by Everitt and Hothorn (2011) is an additional reference for people interested in the execution of multivariate statistical analyses using R, as it offers a handy access of R code for the methods covered in this primer. R code and supporting documentation to accompany the book are all available from the book's website (supplementary resources section). Additional references on the use of R for multivariate analysis are provided in the chapters that follow.

Exercises

1. Locate the data files described in Examples 1.1, 1.3, 1.4., and 1.5, available at the book's website (supplementary resources section). The data sets are provided in two different file formats.

    a. Describe the differences of the two formats. Then use your preferred statistical software to import and visualize variables and number of rows (or sample sizes; see Exercise 3).

    b. **R.** Write and execute R commands to import the files corresponding to Examples 1.1, 1.3, and 1.4. Identify the commands in which option `row.names` = and/or `stringsAsFactors` = *are/are not* necessary. Explain.

2. **R.** Import and create a data frame for the Egyptian skulls data, Example 1.2, using the file "**Egyptian_skulls.csv**". Identify the class of each variable and verify that the class of the numeric variables is `int` (or *integer*). This is a particular numeric subclass of object that R assumes every time our source data are numbers without decimal digits, as integers require less storage space. Now, write and execute R-commands that will allow to select the following subsets of Egyptian skulls:

    a. The first six rows.

    b. The last six rows.

    c. The third column, i.e., the second numeric variable. Is this subset a vector or a data frame?

    d. Variables `Basialveolar_length` and `Nasal_height` for skulls from the Ptolemaic period.

    e. Skulls having `Maximum_breath` greater than 140 mm, displaying on the console `Period` and `Maximum_breath` only.

3. a. Look at any of your univariate statistics textbooks and search for the definitions of the following terms:

       i. Continuous random variable

       ii. Discrete random variable

    iii. Experimental (or observational) unit

    iv. Sample size

b. Characterize the quantitative variables found in Examples 1.1–1.5, by indicating whether they are either continuous or discrete. Also determine the experimental/observational units and the sample sizes for each example.

c. Look again at your textbooks and find the definition and properties of the univariate normal distribution. What type of variable is assumed for a normal distribution: discrete or continuous? What do you intuitively conclude about the types of variables involved in the generalization of the univariate normal distribution, namely, the multivariate normal distribution?

References

Bumpus, H.C. (1898). The elimination of the unfit as illustrated by the introduced sparrow, *Passer domesticus*. *Biological Lectures*, 11th Lecture, pp. 209–226. Woods Hole, MA: Marine Biology Laboratory.

Davies, T.M. (2016). *The Book of R. A First Course in Programming and Statistics*. San Francisco, CA: No Starch Press.

Ekstrøm, C.T. (2016). *The R Primer*. 2nd Edn. Boca Raton, FL: Chapman and Hall.

Everitt, B., and Hothorn, T. (2011). *An Introduction to Applied Multivariate Analysis with R*. New York: Springer.

FAO. (2023). *FAOSTAT. Cost and Affordability of a Healthy Diet (CoAHD)*. www.fao.org/faostat/en/#data/CAHD (Accessed: 27 March 2023).

Herforth, A., Bai, Y., Venkat, A., Mahrt, K., Ebel, A., and Masters, W.A. (2020). *Cost and Affordability of Healthy Diets Across and Within Countries. Background Paper for the State of Food Security and Nutrition in the World 2020*. FAO Agricultural Development Economics Technical Study No. 9. Rome: FAO. https://doi.org/10.4060/cb2431en.

Herforth, A., Venkat, A., Bai, Y., Costlow, L., Holleman, C., and Masters, W.A. (2022). Methods and options to monitor the cost and affordability of a healthy diet globally. Background paper for The State of Food Security and Nutrition in the World 2022. *FAO Agricultural Development Economics Working Paper* 22–03. Rome: FAO.

Highman, C., Kijngam, A., and Manly, B.F.J. (1980). An analysis of prehistoric canid remains from Thailand. *Journal of Archaeological Science* 7: 149–165.

Jones, E., Harden, S., and Crawley, M. (2022). *The R Book*. 3rd Edn. Chichester: Wiley.

Kabacoff, R.I. (2020). *R in Action*. 3rd Edn. New York: Manning.

Long, J.D., and Teetor, P. (2019). *R Cookbook. Proven Recipes for Data Analysis, Statistics and Graphics*. 2nd Edn. Sebastopol: O'Reilly.

McKechnie, S.W., Ehrlich, P.R., and White, R.R. (1975). Population genetics of Euphydryas butterflies. I. Genetic variation and the neutrality hypothesis. *Genetics* 81: 571–594.

Minitab, Inc. (2023). *Minitab 21.4 Statistical Software*. State College, PA. www.minitab.
com.

R Core Team. (2024). *R: A Language and Environment for Statistical Computing*. Vienna:
R Foundation for Statistical Computing. www.R-project.org/.

Thomson, A., and Randall-Maciver, R. (1905). *Ancient Races of the Thebaid*. London:
Oxford University Press.

Venables, W.N., Smith, D.M., and the R Core Team. (2024, February 29). *An Introduction
to R: Notes on R, A Programming Environment for Data Analysis and Graphics,
Version 4.3.3*. https://cran.r-project.org/doc/manuals/R-intro.pdf (Accessed: 11
August 2023).

Zamora Saiz, A., Quesada González, C., Hurtado Gil, L., and Mondéjar Ruiz, D.
(2020). *An Introduction to Data Analysis in R*. Cham: Springer.

2

Matrix Algebra

2.1 The Need for Matrix Algebra

The theory of multivariate statistical methods can only be explained reasonably well with the use of matrix algebra. For this reason, it is helpful, if not essential, to have at least some knowledge of this area of mathematics. This is true even for those who are only interested in using the methods as tools. At first sight, the notation of matrix algebra is certainly somewhat daunting. However, it is not difficult to understand the basic principles, providing that some of the details are accepted on faith.

2.2 Matrices and Vectors

An $m \times n$ *matrix* is an array of numbers with m rows and n columns, considered as a single entity, of the form

$$
\mathbf{A} = \begin{bmatrix} a_{11} & a_{12} & \cdots & a_{1n} \\ a_{21} & a_{22} & \cdots & a_{2n} \\ \vdots & \vdots & \ddots & \vdots \\ a_{m1} & a_{m2} & \cdots & a_{mn} \end{bmatrix}.
$$

The size of a matrix is, at first, a bit counterintuitive: it is referenced by the number of rows and columns. For instance, if $m = 3$ and $n = 2$, then we would say that $\mathbf{A}$ is of size 3×2. This would not be considered the same "size" as a matrix with two rows and three columns (which has size 2×3).

If $m = n$, then it is a *square matrix*. A *column vector* is a matrix with but a single column; for example:

$$
c = \begin{bmatrix} c_1 \\ c_2 \\ \vdots \\ c_m \end{bmatrix}.
$$

DOI: 10.1201/9781003453482-2

If there is only one row, such as $\mathbf{r} = (r_1, r_2, \ldots, r_n)$, we have a *row vector*. Bold type is used to indicate matrices and vectors.

More formally, a vector is defined as an ordered n-tuple of real numbers, i.e., a set of n numbers with a specified order. The n numbers are the coordinates of a point in a n-dimensional Euclidean space, which may be seen as the endpoint of a line segment starting at the origin. Figure 2.1a shows the point (4, 3) in 2-dimensional Euclidean space, while Figure 2.1b displays the vector $\mathbf{v}' = (4, 3)$ as an arrow. The length or *norm* of a two-dimensional vector $\mathbf{v}$ (denoted $\|\mathbf{v}\|$) can be found by the Pythagorean theorem.

For example, the length of $\mathbf{v}' = (4, 3)$ is that of the hypotenuse of a right triangle with base 4 and height 3. Therefore $\|\mathbf{v}'\| = \sqrt{4^2 + 3^2} = 5$.

The *transpose* of a matrix is obtained by interchanging the rows and the columns. Thus, element a_{ij} of the matrix $\mathbf{A}$ will become a_{ji} in its transpose. Also, the transpose of the vector $\mathbf{c}$ is the row vector $\mathbf{c}' = (c_1, c_2, \ldots, c_m)$, and the transpose of the row vector $\mathbf{r}$ is the column vector $\mathbf{r}'$.

Certain special kinds of matrices that are important. A *zero matrix* has all elements equal to zero, so that it is of the form

$$\mathbf{0} = \begin{bmatrix} 0 & 0 & \cdots & 0 \\ 0 & 0 & \cdots & 0 \\ \vdots & \vdots & \ddots & \vdots \\ 0 & 0 & \cdots & 0 \end{bmatrix}.$$

A *diagonal matrix* has zero elements except down the main diagonal:

$$\mathbf{D} = \begin{bmatrix} d_1 & 0 & \cdots & 0 \\ 0 & d_2 & \cdots & 0 \\ \vdots & \vdots & \ddots & \vdots \\ 0 & 0 & \cdots & d_n \end{bmatrix}.$$

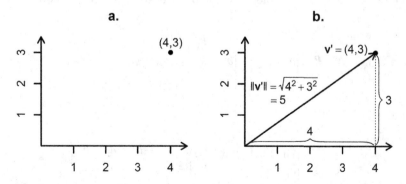

FIGURE 2.1
(a) A point in the two-dimensional Euclidean space. (b) The corresponding two-dimensional vector and interpretation of the norm of a vector.

A *symmetric matrix* is a square matrix that is unchanged when it is transposed, so that $\mathbf{A}' = \mathbf{A}$. An *identity matrix* is a diagonal matrix with all terms on the diagonal equal to one:

$$\mathbf{I} = \begin{bmatrix} 1 & 0 & \cdots & 0 \\ 0 & 1 & \cdots & 0 \\ \vdots & \vdots & \ddots & \vdots \\ 0 & 0 & \cdots & 1 \end{bmatrix}.$$

In matrix algebra, an ordinary number such as 20 is called a *scalar*.

Two matrices are *equal* only if they are the same size and all their elements are equal. For example,

$$\mathbf{A} = \begin{bmatrix} a_{11} & a_{12} & a_{13} \\ a_{21} & a_{22} & a_{23} \\ a_{31} & a_{32} & a_{33} \end{bmatrix} = \begin{bmatrix} b_{11} & b_{12} & b_{13} \\ b_{21} & b_{22} & b_{23} \\ b_{31} & b_{32} & b_{33} \end{bmatrix} = \mathbf{B}$$

only if $a_{11} = b_{11}$, $a_{12} = b_{12}$, $a_{13} = b_{13}$, and so on.

The *trace* of a matrix, defined only for a square matrix, is the sum of its diagonal terms. For example, the trace of the foregoing 3×3 matrix $\mathbf{A}$ has $\text{trace}(\mathbf{A}) = a_{11} + a_{22} + a_{33}$.

2.3 Operations on Matrices

The ordinary arithmetic processes of addition, subtraction, multiplication, and division have their counterparts with matrices. With addition and subtraction, it is just a matter of working element by element with two matrices of the same size. For example, if $\mathbf{A}$ and $\mathbf{B}$ are both of size 3×2, then

$$\mathbf{A} + \mathbf{B} = \begin{bmatrix} a_{11} & a_{12} \\ a_{21} & a_{22} \\ a_{31} & a_{32} \end{bmatrix} + \begin{bmatrix} b_{11} & b_{12} \\ b_{21} & b_{22} \\ b_{31} & b_{32} \end{bmatrix} = \begin{bmatrix} a_{11}+b_{11} & a_{12}+b_{12} \\ a_{21}+b_{21} & a_{22}+b_{22} \\ a_{31}+b_{31} & a_{32}+b_{32} \end{bmatrix}$$

while

$$\mathbf{A} - \mathbf{B} = \begin{bmatrix} a_{11} & a_{12} \\ a_{21} & a_{22} \\ a_{31} & a_{32} \end{bmatrix} - \begin{bmatrix} b_{11} & b_{12} \\ b_{21} & b_{22} \\ b_{31} & b_{32} \end{bmatrix} = \begin{bmatrix} a_{11}-b_{11} & a_{12}-b_{12} \\ a_{21}-b_{21} & a_{22}-b_{22} \\ a_{31}-b_{31} & a_{32}-b_{32} \end{bmatrix}.$$

Multiplication of a matrix **A** by a scalar k is defined by multiplying every element in **A** by k. Thus, if **A** is the 3×2 matrix as shown, then

$$k\mathbf{A} = \begin{bmatrix} ka_{11} & ka_{12} \\ ka_{21} & ka_{22} \\ ka_{31} & ka_{32} \end{bmatrix}.$$

The multiplication of two matrices, denoted by **A·B** or **A** × **B**, is more complicated. To begin with, **A·B** is only defined if the number of columns of the first matrix (**A** here) is equal to the number of rows of the second. Suppose that **A** has size $m \times n$ and **B** having the size $n \times p$. Then, multiplication is defined to produce the result

$$\mathbf{A \cdot B} = \begin{bmatrix} \Sigma a_{1j} \cdot b_{j1} & \Sigma a_{1j} \cdot b_{j2} & \cdots & \Sigma a_{1j} \cdot b_{jp} \\ \Sigma a_{2j} \cdot b_{j1} & \Sigma a_{2j} \cdot b_{j2} & \cdots & \Sigma a_{2j} \cdot b_{jp} \\ \vdots & \vdots & \ddots & \vdots \\ \Sigma a_{mj} \cdot b_{j1} & \Sigma a_{mj} \cdot b_{j2} & \cdots & \Sigma a_{mj} \cdot b_{jp} \end{bmatrix},$$

where the summations are for j from 1 to n. Hence, the element in the ith row and jth column of **A·B** is $\sum a_{ij} \times b_{jk} = a_{i1} \times b_{1k} + a_{i2} \times b_{2k} + \cdots + a_{in} \times b_{nk}$.

When **A** and **B** are both square matrices, then **A·B** and **B·A** are both defined. However, they are not necessarily equal. For example,

$$\begin{bmatrix} 2 & -1 \\ 1 & 1 \end{bmatrix}\begin{bmatrix} 1 & 1 \\ 0 & 1 \end{bmatrix} = \begin{bmatrix} 2\times1-1\times0 & 2\times1-1\times1 \\ 1\times1+1\times0 & 1\times1+1\times1 \end{bmatrix} = \begin{bmatrix} 2 & 1 \\ 1 & 2 \end{bmatrix},$$

whereas

$$\begin{bmatrix} 1 & 1 \\ 0 & 1 \end{bmatrix}\begin{bmatrix} 2 & -1 \\ 1 & 1 \end{bmatrix} = \begin{bmatrix} 1\times2+1\times1 & 1\times(-1)+1\times1 \\ 0\times2+1\times1 & 0\times(-1)+1\times1 \end{bmatrix} = \begin{bmatrix} 3 & 0 \\ 1 & 1 \end{bmatrix}.$$

2.4 Matrix Inversion

The inverse of a scalar k is $k^{-1} = 1/k$. We know that $k \times k^{-1} = 1$. In a similar way, if **A** is a square matrix and $\mathbf{A} \times \mathbf{A}^{-1} = \mathbf{I}$, where **I** is the identity matrix, then $\mathbf{A}^{-1}$ is the inverse of **A**. Notice in the preceding sentence that we said, "if" and "and." Inverses only exist for square matrices, but not all square matrices have inverses (more on which in the following).

As an example, let $\mathbf{A} = \begin{bmatrix} 2 & 1 \\ 1 & 2 \end{bmatrix}$. Then, the inverse matrix for $\mathbf{A}$ is

$$\mathbf{A}^{-1} = \begin{bmatrix} 2 & 1 \\ 1 & 2 \end{bmatrix}^{-1} = \begin{bmatrix} 2/3 & -1/3 \\ -1/3 & 2/3 \end{bmatrix}$$

which can be verified by checking that

$$\mathbf{A}\mathbf{A}^{-1} = \begin{bmatrix} 2 & 1 \\ 1 & 2 \end{bmatrix} \begin{bmatrix} 2/3 & -1/3 \\ -1/3 & 2/3 \end{bmatrix} = \begin{bmatrix} 1 & 0 \\ 0 & 1 \end{bmatrix} = \mathbf{I}$$

Actually, the inverse of a 2×2 matrix, if it exists, can be calculated fairly easily. Let

$$\mathbf{A} = \begin{bmatrix} a & b \\ c & d \end{bmatrix}.$$

The equation for $\mathbf{A}^{-1}$ is

$$\mathbf{A}^{-1} = \begin{bmatrix} a & b \\ c & d \end{bmatrix}^{-1} = \begin{bmatrix} d/\Delta & -b/\Delta \\ -c/\Delta & a/\Delta \end{bmatrix}$$

where $\Delta = a \times d - b \times c$. Here, the scalar Δ is called the *determinant* of the matrix being inverted. Clearly, the inverse is not defined if $\Delta = 0$ because the calculation involves a division by zero. Any square matrix has a determinant, which can be calculated by a generalization of the equation just given for the 2×2 case. If the determinant is zero, then the inverse does not exist, and vice versa. A matrix with a zero determinant is said to be *singular*. For 3×3 and larger matrices, the calculation of the inverse is tedious and best done by using a computer program.

Matrices sometimes arise for which the inverse is equal to the transpose. They are then said to be *orthogonal*. Hence, $\mathbf{A}$ is orthogonal if $\mathbf{A}^{-1} = \mathbf{A}'$.

2.5 Quadratic Forms

Suppose that $\mathbf{A}$ is an n by n matrix and $\mathbf{x}$ is a column vector of length n. Then, the quantity $Q = \mathbf{x}'\mathbf{A}\mathbf{x}$ is a scalar that is called a *quadratic form*. This scalar can also be expressed as

$$Q = \sum_{i=1}^{n} \sum_{j=1}^{n} x_i a_{ij} x_j.$$

2.6 Eigenvalues and Eigenvectors

Consider the set of linear equations

$$a_{11}x_1 + a_{12}x_2 + \cdots + a_{1n}x_n = \lambda x_1$$
$$a_{21}x_1 + a_{22}x_2 + \cdots + a_{2n}x_n = \lambda x_2$$
$$\vdots$$
$$a_{n1}x_1 + a_{n2}x_2 + \cdots + a_{nn}x_n = \lambda x_n$$

where λ is a scalar. These can also be written in matrix form as $\mathbf{Ax} = \lambda x$ or $(\mathbf{A} - \lambda\mathbf{I})x = 0$, where $\mathbf{I}$ is the $n \times n$ identity matrix and 0 is an $n \times 1$ vector of zeros. These equations only hold for certain particular values of λ, which are called the *latent roots* or *eigenvalues* of $\mathbf{A}$. There can be up to n of these eigenvalues. Given the ith eigenvalue λ_i, the equations can be solved by arbitrarily setting $x_1 = 1$, and the resulting vector of x values with transpose $x' = (1, x_2, \ldots, x_n)$, or any multiple of this vector, is called the ith latent root or the ith eigenvector of the matrix $\mathbf{A}$. Also, the sum of the eigenvalues is equal to the trace of $\mathbf{A}$ defined in Section 2.2 so that trace($\mathbf{A}$) = $\lambda_1 + \lambda_2 + \cdots + \lambda_n$.

2.7 Vectors of Means and Covariance Matrices

Population and sample values for a single random variable are often summarized by the values for the mean and variance. Thus, if a sample of size n yields the values $x_1, x_2, \ldots, x_n$, then the sample mean is defined to be

$$\bar{x} = x_1 + x_2 + \cdots + x_n = \sum_{i=1}^{n} x_i$$

while the sample variance is

$$s^2 = \sum_{i=1}^{n} (x_i - \bar{x})^2 / (n-1).$$

These are estimates of the corresponding population parameters, which are the population mean μ and the population variance σ^2.

In a similar way, multivariate populations and samples can be summarized by mean vectors and covariance matrices. Suppose that there are p variables $X_1, X_2, \ldots, X_p$ being considered, and that a sample of n values for each of

these variables is available. Let the sample mean and sample variance for the
jth variable be $\bar{x}_j$ and s_j^2, respectively. In addition, define the *sample covariance*
between variables X_j and X_k by

$$c_{jk} = \sum_{i=1}^{n}(x_{ij} - \bar{x}_j)(x_{ik} - \bar{x}_k)/(n-1)$$

where x_{ij} is the value of variable X_j for the ith multivariate observation. This
covariance is then a measure of the extent to which there is a linear relationship between X_j and X_k, with a positive value indicating that large values of
X_j and X_k tend to occur together, and a negative value indicating that large
values for one variable tend to occur with small values for the other variable.
It is related to the ordinary correlation coefficient between the two variables,
which is defined to be

$$r_{jk} = c_{jk} / (s_j s_k).$$

Here s_j is the sample standard deviation; it is equal to the positive square
root of the sample variance. The definitions imply that $c_{kj} = c_{jk}$, $r_{kj} = r_{jk}$, $c_{jj} = s^2$,
and $r_{jj} = 1$.

With these definitions, the transpose of the sample mean vector is
$x' = (\bar{x}_1, \bar{x}_2, ..., \bar{x}_p)$, which is an estimate of the transpose of the population
vector of means $\mu' = (\mu_1, \mu_2, ..., \mu_p)$.

Furthermore, the sample matrix of variances and covariances, or the
covariance matrix, is

$$C = \begin{bmatrix} c_{11} & c_{12} & \cdots & c_{1p} \\ c_{21} & c_{22} & \cdots & c_{2p} \\ \vdots & \vdots & \ddots & \vdots \\ c_{p1} & c_{p2} & \cdots & c_{pp} \end{bmatrix}$$

where $c_{ii} = s^2$. This is also sometimes called the *sample dispersion matrix*, and it
measures the amount of variation in the sample as well as the extent to which
the p variables are correlated. It is an estimate of the population covariance
matrix

$$\Sigma = \begin{bmatrix} \sigma_{11} & \sigma_{12} & \cdots & \sigma_{1p} \\ \sigma_{21} & \sigma_{22} & \cdots & \sigma_{2p} \\ \vdots & \vdots & \ddots & \vdots \\ \sigma_{p1} & \sigma_{p2} & \cdots & \sigma_{pp} \end{bmatrix}.$$

Finally, the sample correlation matrix is

$$\mathbf{R} = \begin{bmatrix} 1 & r_{12} & \cdots & r_{1p} \\ r_{21} & 1 & \cdots & r_{2p} \\ \vdots & \vdots & \ddots & \vdots \\ r_{p1} & r_{p2} & \cdots & 1 \end{bmatrix}.$$

Again, this is an estimate of the corresponding *population correlation matrix*. An important result for some analyses is that if the observations for each of the variables are coded by subtracting the sample mean and dividing by the sample standard deviation, then the coded values will have a mean of zero and a standard deviation of one for each variable. In that case, the sample covariance matrix will equal the sample correlation matrix, that is, $\mathbf{C} = \mathbf{R}$.

2.8 Missing Values in Matrices Used for Multivariate Statistics

Missing values can cause more problems with multivariate data than with univariate data. When there are many variables being measured on each individual, it is often the case that one or two of these variables have missing values. If individuals with any missing values are excluded from an analysis, a large proportion of individuals would be excluded, which may be completely impractical. For example, in the study of ancient human populations, skeletons are frequently broken and incomplete.

Texts on multivariate analysis are often quite silent on the question of missing values. To some extent, this is because doing something about missing values is by no means a straightforward matter. Various solutions can be tried in the presence of missing data, but those solutions may alter the statistical reliability of results by the simple fact that matrix operations are based on incomplete data. If data completeness is the property sought (i.e., all variables of interest are measured for every sampling unit), but some sampling units are incomplete, the solution is simply to delete as many variables or as many objects where any of them are missing. However, this procedure usually leads to the unwanted effect of a sample size decrease, which can be remarkable serious as to warrant a valid statistical analysis.

An intermediate solution to the missing data problem is often applicable to the computation of the correlation or covariance between each pair of variables, in which all complete pairs of observations on those variables are used. This procedure is called the *pairwise-complete observations method* for handling missing data or the *available-case analysis* (Everitt and Hothorn,

2011). However, this method may affect some multivariate analyses as the resulting covariance or correlation matrices may not be positive semi-definite, a concept defined in terms of quadratic forms (for a definition, see Fieller, 2021). The positive semi-definite property in covariance/correlation matrices is helpful for the application of methods like PCA (Chapter 6) where eigenvalues are computed and interpreted as fractional variances of index variables contributing to the overall variance in a multivariate data set. If a matrix is positive semi-definite, their corresponding eigenvalues are nonnegative, an attribute that variances are expected to have.

Another strategy to cope with missing data is imputation, meaning that the unknown observations are replaced by suitable estimates, using an appropriate algorithm based on the nature of the data. In practice, computer packages sometimes include a facility for data imputation by various methods of varying complexity. This procedure will work satisfactorily provided that only a small proportion of values are missing.

For a detailed discussion of methods for dealing with missing data, see the texts by Little and Rubin (2020), and van Buuren (2018).

2.9 Further Reading

This short introduction to matrix algebra will help to understand the methods described in the remainder of this book and some of the theory behind these methods. However, for a better understanding of the theory, more knowledge and proficiency are required. There are many books of various lengths that cover what is needed just for statistical applications. Four of these are by Fieller (2021), Searle and Khuri (2017), Gentle (2017), and Banerjee and Roy (2014). A web search on the topic of matrix algebra should yield some useful free books and course notes. For R users interested in learning matrix algebra and applications to statistical methods, the first three books referred to earlier are strongly recommended.

Exercises

1. **R.** Verify numerically the orthogonality (Section 2.4) of

$$\mathbf{A} = \begin{bmatrix} 3/5 & 4/5 \\ -4/5 & 3/5 \end{bmatrix}.$$

2. **R.** Let $\mathbf{A} = \begin{bmatrix} 1 & 3 \\ 0 & 1 \end{bmatrix}$. Calculate: $\mathbf{A}\cdot\mathbf{A} = \mathbf{A}^2$, $\mathbf{A}\cdot\mathbf{A}\cdot\mathbf{A} = \mathbf{A}^3$ in two ways:

   a. using the `%*%` operator and

b. using the `%^%` operator from the `expm` package (Maechler et al., 2023).

c. By induction, determine a general expression of A^n, for any positive integer n.

d. Given a real number m, generalize the results in part c) for any matrix.

$$A_m = \begin{bmatrix} 1 & m \\ 0 & 1 \end{bmatrix}.$$

3. *Manual calculation of eigenvalues and eigenvectors.* The equation $(A - \lambda I)x = 0$ described in Section 2.6 is crucial for the calculation of eigenvalues and eigenvectors of any $n \times n$ matrix A and any nonzero column vector x. The theory of linear equations states that a nonzero solution x exists if and only if the determinant of $(A - \lambda I)$ is zero. This implies that any eigenvalue λ of A satisfies the so-called *characteristic equation* of A:

$$\det(A - \lambda I) = 0.$$

$$\text{Let } A = \begin{bmatrix} 1 & 4 \\ 2 & 3 \end{bmatrix}. \text{ In this case, } \lambda I = \begin{bmatrix} \lambda & 0 \\ 0 & \lambda \end{bmatrix}.$$

a. Find $(A - \lambda I)$ and then obtain $\det(A - \lambda I)$ as a function of λ, using the expression given in section 2.4 for the determinant Δ of a 2×2 matrix.

b. Solve the (quadratic) characteristic equation for λ. Denote the roots as λ_1 and λ_2.

c. Substitute the value of λ_1 in the expression for $(A - \lambda I)$ established in (a), and let $x' = (x_1, x_2) \neq 0$, where $0' = (0,0)$. Verify that the two equations in terms of x_1 and x_2 after substitution of λ_1 in $(A - \lambda_1 I)x = 0$ reduce to one equation only. This implies that there are infinite eigenvectors x_1 associated to the eigenvalue λ_1.

d. To find *an* eigenvector x associated to the eigenvalue λ_i, it was suggested (Section 2.6) to arbitrarily set $x_1 = 1$. Use the equations obtained in (c) and let $x_1 = 1$ to get a numeric eigenvector of A associated to λ_1.

e. Redo parts (c) and (d), but now using the second eigenvalue λ_2 and setting $x_1 = 1$.

f. **R**. Compute the eigenvalues of A using the function eigen. Check these results match with those in part (b).

g. **R**. The eigen function stores eigenvectors in the object `eigen(A)$vectors`. Moreover, the eigenvector associated to the largest eigenvalue of **A** is the first column in that object, that can be invoked by writing `eigen(A)$vectors[,1]`. This eigenvector is different from that found in part (c). However, they are multiples of each other. Multiply the first eigenvector by the inverse of its first element, (`1/eigen(A)$vectors[1,1]`), to obtain the same eigenvector found in part (d). Do the same for the second eigenvector stored in `eigen(A)$vectors[,2]`, by multiplying it by its first element (`1/eigen(A)$vectors[1,2]`). Compare with part (e).

4. In the United States, citizens are eligible to vote in the general election for any candidate from any party. It does not matter if voters are registered with a political party or who they voted for in the past. Assume that a person who voted for Democrats in the last election has a probability of 0.8 to vote for that party again, while the probability that a person who voted for Republicans in the last general election will vote again for them is 0.7. Here we have two "states": Democrat and Republican. Figure 2.2 shows the transitions between states (i.e., between the last and the next elections).

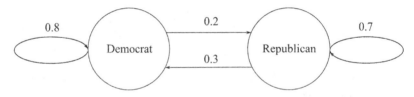

FIGURE 2.2
State transition diagram in Exercise 3.

In tabular form:

		Next election	
		Democrat (D)	**Republican (R)**
Last election	Democrat (D)	0.8	0.2
	Republican (R)	0.3	0.7

Or, more succinctly, as a 2×2 *transition matrix*

$$\mathbf{M} = \begin{array}{c} \\ D \\ R \end{array} \begin{array}{cc} D & R \\ \begin{bmatrix} 0.8 & 0.2 \\ 0.3 & 0.7 \end{bmatrix} \end{array}.$$

This transition matrix shows the probabilities of potential changes between states at two consecutive elections. To represent the distribution among the states at present or future elections, we use a row vector of probabilities called a *state vector*, v_n; it has one column for each state, showing the distribution by state at a given point in time, n. All entries are between 0 and 1 inclusive, and their sum is 1. Assume that the *initial state vector* of party preferences is evenly divided: $v_0 = (0.5, 0.5)$. This exercise exemplifies a *discrete Markov chain*, usually considered as a model of the evolution of "states" in probabilistic systems (see Searle and Khuri, 2017, Chapter 2).

a. Determine the distribution of voters after one and two transitions, i.e., for the next election and for the election after the next, by computing $v_1 = v_0 M$ and $v_2 = v_1 M$.

b. Express v_2 in terms of v_0 and M only. Make use of the notation: $M \cdot M = M^2$ (see Exercise 2) and compute v_2.

c. Compute M^3, and compute v_3 in terms of v_0 and M only. Write a general expression for v_n, the state vector for n transitions, in terms of v_0 and M.

d. **R**. To compute matrix powers of the form M^n in R, the %^% operator implemented in the expm package (Maechler et al., 2023) is quite useful. Examine M^n for $n > 3$, and verify that

$$\lim_{n \to \infty} M^n = E = \begin{bmatrix} 0.6 & 0.4 \\ 0.6 & 0.4 \end{bmatrix}.$$

e. Compute the *limiting distribution* of the Markov chain:

$$v_\infty = \lim_{n \to \infty} v_n = \lim_{n \to \infty} v_0 M^n = v_0 E.$$

Does the vector v_∞ change if $v_0 = (0.5, 0.5)$ is substituted by a row vector of the form $v_0 = (\pi, 1 - \pi)$, with $0 \le \pi \le 1$? Give a conclusion about the dependence or independence of the limiting distribution on the initial probabilities $v_0 = (\pi, 1 - \pi)$.

f. **R**. Verify that $\lambda = 1$ is an eigenvalue of M. Show that the initial vector state $v_0 = (0.5, 0.5)$ is also an eigenvector associated to the eigenvalue $\lambda = 1$.

References

Banerjee, S., and Roy, A. (2014). *Linear Algebra and Matrix Analysis for Statistics*. Boca Raton, FL: Chapman and Hall/CRC.

Fieller, N. (2021). *Basics of Matrix Algebra for Statistics with R*. Boca Raton, FL: Chapman and Hall/CRC.

Gentle, J.E. (2017). *Matrix Algebra. Theory, Computations and Applications in Statistics*. Cham: Springer.

Little, R.A., and Rubin, D.B. (2020). *Statistical Analysis with Missing Data*. 2nd Edn. New York: Wiley.

Maechler, M., Dutang, C., and Goulet, V. (2023). *expm: Matrix Exponential, Log, 'etc'*. R package version 0.999-7. https://CRAN.R-project.org/package=expm.

Searle, S.R., and Khuri, A.I. (2017). *Matrix Algebra Useful for Statistics*. 2nd Edn. Hoboken, NJ: Wiley.

van Buuren, S. (2018). *Flexible Imputation of Missing Data*. 2nd Edn. Boca Raton, FL: CRC Press.

Appendix: Matrix Algebra in R

A summary of the main R functions useful for handling matrices and performing matrix operations is given here. A more detailed description and illustrative examples about their usage, as well as additional matrix functions, are given in the "Supplementary Resources" repository. The complete set of options for each function can be found in the corresponding R help documents. Here, **x** is a vector, and **A** and **B** are matrices.

Function	Description	Useful options / Comments
matrix(x, nrow, ncol, ...)	Generates a matrix column-wise, with nrow rows and ncol columns	byrow = TRUE generates a matrix row-wise
cbind(x1, x2, ..., xp)	Binds p vectors x1, x2, ..., xp by column, each vector of length n, to produce a n × p matrix	rbind(x1, x2, ..., xn) binds n vectors x1, x2, ..., xn by row, each vector of length p, to produce a n × p matrix
dim(**A**), nrow(**A**), ncol(**A**)	Matrix dimension	
t(**A**)	Transpose of a matrix	
matrix(0, n, m)	The n × m zero matrix	
diag(c(s1, s2, ..., sn))	Diagonal matrix with scalars s1, s2, ..., sn in the main diagonal. The output is a square matrix of dimension n	diag(s), with s an integer scalar, produces an identity matrix of dimension s
diag(**A**)	Diagonal extraction of the square matrix **A**	
sum(diag(**A**))	Trace of a square matrix **A**	
A + **B**, **A** − **B**	Addition and subtraction of conformable matrices	
k * **A**	Multiplication of a matrix by a scalar k	**A** * **B** produces an element-wise multiplication of two conformable matrices. This is not the standard matrix multiplication
A %*% **B**	Matrix multiplication	
det(**A**)	Determinant of a square matrix	
solve(**A**)	Inverse of a square matrix	solve(**A**, **B**) is the solution x of the simultaneous system of linear equations **A** %*% **x** = **B**; **B** usually is a column vector.

(Continued)

(Continued)

`eigen(`**A**`)`	Spectral decomposition of the input matrix. A list with two elements: one vector of eigenvalues, and a matrix of eigenvectors arranged in columns of norm 1 each	The eigenvalues and the eigenvectors are invoked as `eigen(`**A**`)$values` and `eigen(`**A**`)$vectors`, respectively
`colMeans(`**A**`)`, `rowMeans(`**A**`)`	Column or row means	Related functions: `colSums(`**A**`)`, `rowSums(`**A**`)`
`cov(`**A**`)`	Given a numeric n × p matrix **A**, creates a p × p variance-covariance matrix	Alternatively, **A** can be a (numeric) data frame
`cor(`**A**`)`	Given a numeric n × p matrix **A**, creates a p × p variance-correlation matrix	Alternatively, **A** can be a (numeric) data frame

The functions listed, and a few other basic matrix functions not considered in Chapter 2, are available from the default base package. Other useful matrix functions have been implemented in additional packages. Two of those packages are `matlib` (Friendly et al., 2022) designed for tutorial purposes in teaching and learning matrix algebra ideas and applications to statistical methods, and `Matrix` (Bates et al., 2024), containing a more advanced collection of matrix functions.

References

Bates, D., Maechler, M., and Jagan, M. (2024). *Matrix: Sparse and Dense Matrix Classes and Methods.* R package version 1.7-0. https://CRAN.R-project.org/package=Matrix.

Friendly, M., Fox, J., and Chalmers, P. (2022). *matlib: Matrix Functions for Teaching and Learning Linear Algebra and Multivariate Statistics.* R package version 0.9.6. https://CRAN.R-project.org/package=matlib.

3

Displaying Multivariate Data

3.1 The Problem of Displaying Many Variables in Two Dimensions

Graphs must be displayed in two dimensions, either on paper or on a computer screen. It is, therefore, straightforward to show one variable plotted on a vertical axis against a second variable plotted on a horizontal axis. For example, Figure 3.1 shows the alar extent plotted against the total length for the 49 female sparrows measured by Hermon Bumpus in his study of natural selection, which has been described in Example 1.1. Such plots allow one or more other characteristics of the objects being studied to be shown as well, and in the case of Bumpus' sparrows, survival and nonsurvival are

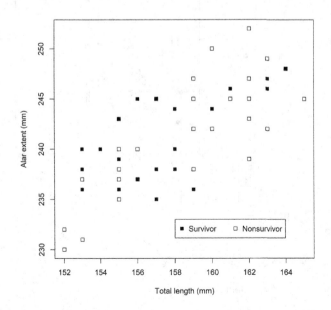

FIGURE 3.1
Alar extent plotted against total length for the 49 female sparrows measured by Hermon Bumpus.

DOI: 10.1201/9781003453482-3

indicated. These plots are simple and can be produced in Excel or another spreadsheet program, as well as in all standard statistical packages. The plots can also be produced using R code, provided in the Section A4.4 appendix and in the supplementary materials section, for both this type of plot and others described here.

It is considerably more complicated to show one variable plotted against another two, but still possible. Thus, Figure 3.2 shows beak and head lengths plotted against total lengths and alar lengths for the 49 sparrows. Again, different symbols are used for survivors and nonsurvivors. This is called a *three-dimensional (3D) plot* and can be produced in many standard statistical packages.

It is not possible to show one variable plotted against another three at the same time in some extension of a 3D plot. Hence, there is a major problem in showing in a simple way the relationships that exist between the individual objects in a multivariate set of data where those objects are each described by four or more variables. Various solutions to this problem have been proposed and are discussed in this chapter.

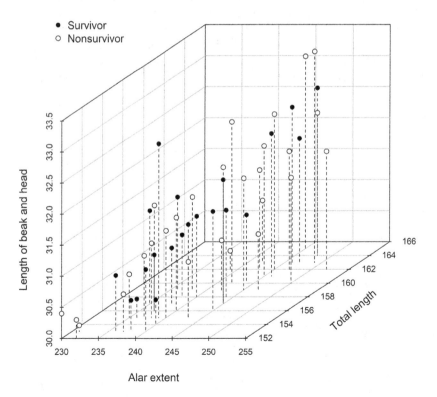

FIGURE 3.2
Length of the beak and head plotted against the total length and alar extent (all in millimeters) for the 49 female sparrows measured by Hermon Bumpus.

3.2 Plotting Index Variables

One approach to making a graphical summary of the differences between objects that are described by more than four variables involves plotting the objects against the values of two or three index variables. Indeed, a major objective of many multivariate analyses is to produce index variables that can be used for this purpose, a process that is sometimes called *ordination*. For example, principal components, as discussed in Chapter 6, provide one type of index variables. A plot of the values of Principal Component 2 against the values of Principal Component 1 can then be used as a means of representing the relationships between objects graphically, and a display of Principal Component 3 against the first two principal components can also be used if necessary.

The use of suitable index variables has the advantage of reducing the problem of plotting many variables to two or three dimensions but has the potential disadvantage that some key difference between the objects may be lost in the reduction. This approach is discussed in various different contexts in the chapters that follow and will not be considered further here.

3.3 The Draftsman's Plot

A draftsman's plot, also called a *scatterplot matrix*, of multivariate data consists of plots of the values for each variable against the values for each of the other variables, with the individual graphs being small enough that they can all be viewed at the same time. This has the advantage of only needing two-dimensional plots but has the disadvantage that some aspect of the data that is only apparent when three or more variables are considered together will not be apparent.

An example is shown in Figure 3.3. Here, the five variables measured by Hermon Bumpus on 49 sparrows (total length, alar extent, length of beak and head, length of humerus, and length of the keel of the sternum, all in millimeters) are plotted for the data given in Table 1.1. Different symbols are used for the measurements on survivors (birds 1–21) and nonsurvivors (birds 22–49). Regression lines are also sometimes added to the plots.

This type of plot is obviously good for showing the relationships between pairs of variables and highlighting the existence of any objects that have unusual values for one or two variables. It can, therefore, be recommended as part of many multivariate analyses and is available in many statistical packages and using R code. Some packages and R also allow the option of specifying the horizontal and vertical variables without insisting that these are the same.

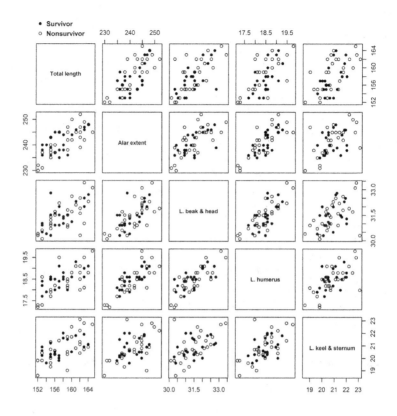

FIGURE 3.3
Draftsman's plot (scatterplot matrix) of the bird number and five variables measured on 49 female sparrows, with points distinguished according to survivorship status. The variables are the total length, the alar extent, the length of the beak and head, the length of the humerus, and the length of the keel of the sternum, with obvious abbreviations, all in millimeters.

The individual objects are not easily identified on a draftsman's plot, and it is, therefore, usually not immediately clear which objects are similar and which are different. Therefore, this type of plot is not suitable for show-ing relationships between objects, as distinct from relationships between variables.

3.4 The Representation of Individual Data Points

An approach to displaying data that is more truly multivariate involves representing each of the objects for which variables are measured by a symbol, with different characteristics of this symbol varying according to different variables. A number of different types of symbol have been proposed for this purpose, including faces (Chernoff, 1973) and stars (Welsch, 1976).

Consider the data in Table 1.4 on mean values of six mandible measurements for seven canine groups, as discussed in Example 1.4. Here, an important question concerns which of the other groups is most similar to the prehistoric Thai dog, and it can be hoped that this will become apparent from a graphical comparison of the groups. To this end, Figure 3.4 shows the data represented by faces and stars, respectively.

For the faces, there was the following connection between features and the variables: mandible breadth to height of face, height of eyes and width

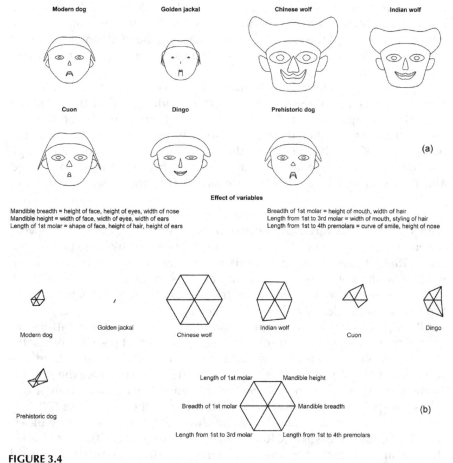

FIGURE 3.4
Graphical representation of mandible measurements on different canine groups using (a) Chernoff faces and (b) stars. See text for details.

of nose; mandible height to width of face, width of eyes, and width of ears; length of first molar to shape of face, height of hair, and height of ears; breadth of first molar to height of mouth and width of hair; length from first to third molar to width of mouth and styling of hair; and length from first to fourth premolars to the amount of smile and height of nose. For example, the height of face is largest for the Chinese wolf, with the maximum mandible breadth of 13.5 mm, and smallest for the golden jackal, with the minimum mandible length of 8.1 mm. It is apparent from the plots that prehistoric Thai dogs are most similar to modern Thai dogs and most different from Chinese and Indian wolves.

For the stars, the six variables were assigned to rays in the order (X_1) mandible breadth, (X_2) mandible height, (X_3) length of first molar, (X_4) breadth of first molar, (X_5) length from first to third molar, and (X_6) length from first to fourth premolars. The mandible breadth is represented by the ray corresponding to three o'clock, and the other variables follow in a counterclockwise order, as indicated by the key that accompanies the figure. Inspection of the stars indicates again that the prehistoric Thai dogs are most similar to modern Thai dogs and most different from Chinese and Indian wolves.

Suggestions for alternatives to faces and stars, and a discussion of the relative merits of different symbols, are provided by Everitt (1978), Toit et al. (1986, chapter 4) and Yau (2011). In summary, it can be said that the use of symbols has the advantage of displaying all variables simultaneously, but the disadvantage that the impression gained from the graph may depend quite strongly on the order in which objects are displayed and the order in which variables are assigned to the different aspects of the symbol.

The assignment of variables is likely to have more effect with faces than with stars, because variation in different features of the face may have very different impacts on the observer, whereas this is less likely to be the case with different rays of a star. For this reason, the recommendation is often made that alternative assignments of variables to features should be tried with faces to find what seems to be the best. The subjective nature of this type of process is clearly rather unsatisfactory.

Although the use of faces, stars, and other similar representations for the values of variables on the objects being considered seems to be useful under some circumstances, the fact is that this is seldom done. One reason is the difficulty in finding computer software to produce the graphs. In the past, this software was reasonably easily available. Barely a few statistical packages that have endured all over the years, like SYSTAT (SYSTAT Software Inc., 2017), Statgraphics (Statgraphics Technologies, Inc., 2023), and Statistica (Cloud Software Group, Inc., 2024), still offer those graphs; regrettably, they are harder to find in newest statistical software.

3.5 Profiles of Variables

Another way to represent objects described by several variables that are measured on them is by lines that show the profile of variable values. A simple
way to draw these involves just plotting the values for the variables, as shown
in Figure 3.5 for the seven canine groups that have already been considered.
The similarity between prehistoric and modern Thai dogs noted from the
earlier graphs is still apparent, as is the difference between prehistoric dogs
and Chinese wolves. In this graph, the variables have been plotted in order of
their average values for the seven groups to help in emphasizing similarities
and differences.

An alternative representation, using bars instead of lines, is shown in
Figure 3.6. Here, the variables are in their original order because there seems
little need to change this when bars are used. The conclusion about similarities and differences between the canine groups is exactly the same as derived
from Figure 3.5.

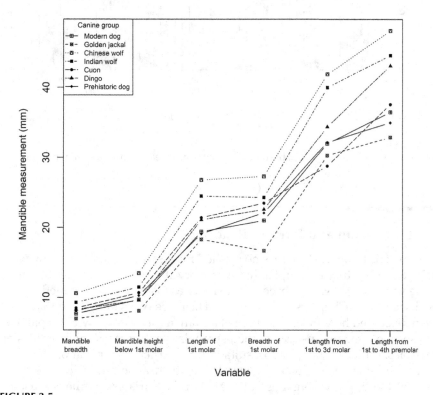

FIGURE 3.5
Profiles of variables for mandible measurements on seven canine groups. The variables are in
order of increasing average values.

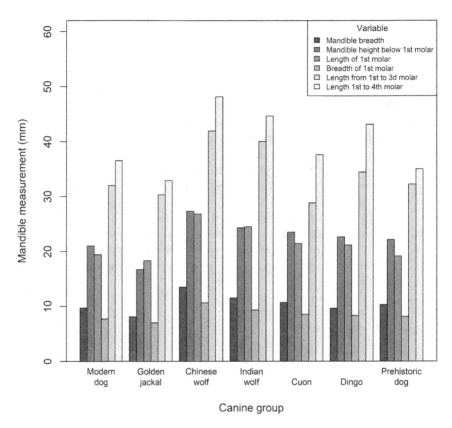

FIGURE 3.6
An alternative way to show variable profiles, using bars instead of the lines used in Figure 3.5.

3.6 Discussion and Further Reading

It seems fair to say that there is no method for displaying data on many variables at a time that is completely satisfactory if it is not desirable to reduce these variables to two or three index variables (using one of the methods to be discussed in Chapters 6 through 12). The three types of method that have been discussed here involve the use of a draftsman's display with all pairs of variables plotted against each other, symbols (stars or faces), and profiles of variables. Which of these is most suitable for a particular application depends on the circumstances, but as a general rule, the draftsman's display is good for highlighting relationships between pairs of variables, while the use of symbols or profiles is good for highlighting unusual cases and similar cases.

For further information about the theory of the construction of graphs in general, see the books by Chambers et al. (1983), Cleveland (1994), Tufte

(2001), and Mazza (2009). More details on graphical methods specifically for multivariate data are in the books by Toit et al. (1986), Jacoby (1999), Mazza (2009), and Yau (2011).

Exercises

1. **R**. Access the data about the cost and affordability of a healthy diet in Europe, Example 1.5. Use the function implemented in "**scatter. grid3d.R**" to construct a 3D scatterplot for CSTS (cost of starchy staples), CASF (cost of animal source food) and COFT (cost of oil and fat). Identify countries (points) according to their EU/non-EU status. Which group of countries appear to pay more for these three types of food?

2. **R**. Construct a draftsman's plot for the skulls data (Example 1.2), using the `pairs` function. Identify points by `Period` employing different point characters and colors. Is there any linear or curvilinear trend between pairs of variables? Comment.

3. **R**. Construct Chernoff faces for the ten variables measured in 16 colonies of the butterfly *Euphydryas editha* (Example 1.3). Use the `faces` function from package `aplpack`, in a 2 × 8 array of faces. Redo the Chernoff faces with `faces2` function from package `TeachingDemos` (Snow, 2024), or with function `PlotFaces` from package `Desctools` (Signorell, 2024). Are there variables that markedly determine face features? Explain.

4. **R**. Build a star plot for all variables of the butterfly data, using the `stars` function. After this, modify the call to the `stars` function by adding option `draw.segments = TRUE` to construct an alternative multivariate plot using colored segments. Which diagram is more efficient in highlighting differences between butterfly colonies: Chernoff faces, stars, or segments?

5. **R**. Create a six-variable subset of the butterfly data containing the six different types of phosphoglucose-isomerase (*Pgi*) mobility genes only. Represent the 16 colonies based on these genetic variables in two ways, lines, and bars, and describe any similarities between colonies.

References

Chambers, J.M., Cleveland, W.S., Kleiner, B., and Tukey, P.A. (1983). *Graphical Methods for Data Analysis*. Belmont, CA: Wadsworth.

Chernoff, H. (1973). Using faces to represent points in K-dimensional space graphically. *Journal of the American Statistical Association* 68: 361–368.

Cleveland, W.S. (1994). *The Elements of Graphing Data*. Rev. Edn. Summit, NJ: Hobart.

Cloud Software Group, Inc. (2024). *Spotfire Statistica®*. Version 14.2.0. Palo Alto, CA.

Everitt, B. (1978). *Graphical Techniques for Multivariate Data*. New York: North-Holland.

Jacoby, W.G. (1999). *Statistical Graphics for Visualizing Multivariate Data*. Thousand Oaks, CA: SAGE.

Mazza, R. (2009). *Introduction to Information Visualization*. London: Springer.

Signorell, A. (2024). *DescTools: Tools for Descriptive Statistics*. R package version 0.99.54. https://CRAN.R-project.org/package=DescTools.

Snow, G. (2024). *TeachingDemos: Demonstrations for Teaching and Learning*. R package version 2.13. https://CRAN.R-project.org/package=TeachingDemos.

Statgraphics Technologies, Inc. (2023). *Statgraphics Centurion 19*. The Plains, VA.

SYSTAT Software Inc. (2017). *Systat Version 13.2 for Windows*. Palo Alto, CA.

Toit, S.H.C., Steyn, A.G.W., and Stumf, R.H. (1986). *Graphical Exploratory Data Analysis*. New York: Springer.

Tufte, E.R. (2001). *The Visual Display of Quantitative Information*. 2nd Edn. Cheshire, CT: Graphics.

Welsch, R.E. (1976). Graphics for data analysis. *Computers and Graphics* 2: 31–37.

Yau, N. (2011). *Visualize This: The Flowing Data Guide to Design, Visualization, and Statistics*. Indianapolis, IN: Wiley.

Appendix: Displaying Multivariate Data in R

A.1 Two-Dimensional Scatterplots

These are easily produced with `plot`, a function introduced in the appendix of Section A4.4. Points in a scatterplot can be identified according to a factor variable by adding a point character option of the `plot` function, namely, `pch`. This is a commonly used graphical parameter indicating the plotting character or symbol for the points, either a single character or an integer code for one of a set of graphics symbols. A full list of `pch` codes can be found in the help document of the *low-level plotting command* `points`; the default code is 1, an empty circle. If a single scalar is chosen for `pch` (e.g., `pch=1`), its effect is vectorized; i.e., it will affect the appearance of all points indicated in the `plot` function. As an example, the data set **Bumpus_sparrows.csv** contains the factor `Survivorship`, which distinguishes nonsurvivor and survivor sparrows. To implement the identification of points in the scatterplot of the `Total_length` against the `Alar_extent` according to survivorship status (as shown in Figure 1.3), conditioning is necessary:

```
pch = ifelse(Survivorship == "S", 15, 0)
```

`ifelse` tests the vector `Survivorship`, element by element, and evaluates whether the level of the factor `Survivorship` is "S". If this is TRUE, `pch` takes the value of the second argument for that point (15, a filled square); otherwise, the point will be displayed using a pch code given by the third argument (0, an open square). Supplementary material for Chapter 3 gives the full annotated code of this strategy of distinguishing points for Figure 3.1.

A.2 Three-Dimensional Scatterplots

The `scatterplot3d` package (Ligges and Mächler, 2003) implements a function with the same name for 3D-plotting. The simplest call to this function is

```
scatterplot3d(x, y, z, ...)
```

where the arguments refer to the variables to be plotted on the *x*, *y*, and *z* axes, respectively. Point identification is also possible, in a similar fashion to that described, for a two-dimensional scatterplot using `plot`. Figure 3.2 was produced with function `scatter.grid3d`, an improved version of `scatterplot3d`, to allow additional decorations in the plot representation. The function `scatter.grid3d` can be found in the source file **scatter.grid3d.R**, downloadable from the book's website, and the supplementary material for Chapter 3 gives the annotated full code for this example.

Another variation of `scatterplot3d` has been implemented in the `car` package (Fox and Weisberg, 2019), offering the `scatter3d` function. Moreover, `scatter3d` depends on the `plot3d` function included in the `rgl` package (Murdoch and Adler, 2024), a tool allowing 2D/3D interactive graphics based on OpenGL, a cross-language, cross-platform API for the representation of 2D and 3D vector graphics.

A.3 Draftsman's Plots

R has a way to produce a matrix of scatterplots using a basic command called `pairs`. As an example, in Figure 3.3, a draftsman's plot for the Bumpus' sparrows data, the variable names are shown on the main diagonal, and the points are identified by their `Survivorship` status. This figure can be recreated using the full code available in the supplementary material for Chapter 4. The user may want to display histograms, boxplots, density plots, etc., on the plot's diagonal, but some programming skills are needed. There are several ways to circumvent the programming hassles, all of them involving R packages. Thus, the `SciViews` package (Grosjean, 2024) offers the inclusion of the argument `diag.panel` to the function `pairs`, as an easy way to specify the diagonal elements (e.g., `diag.panel = "hist"` for histograms, `diag.panel = "boxplot"`, etc.). Another friendlier function allowing customized draftsman's plots is `scatterplotMatrix`, present in the `car` package (Fox and Weisberg, 2019). More sophisticated functions for similar purposes are given in the packages `lattice` (Sarkar, 2008) and `Ggally` (Schloerke et al., 2024). See the supplementary material for instructions about accessing example R code with all these functions.

A.4 Representation of Individual Data Points: Chernoff Faces and Stars

Three R packages that provide functions to make Chernoff faces are `aplpack` (Wolf, 2019), `DescTools` (Signorell, 2024), and `TeachingDemos`

(Snow, 2024). Through the single function `faces`, the `aplpack` package allows complex displays (e.g., colored face regions, scatterplots of two variables (X, Y) where faces are placed in the (X, Y) coordinates, etc.). In contrast, the very basic `PlotFaces` function offered by `DescTools` is based on a simpler implementation of `faces`. Figure 3.4b was produced using `faces` in `aplpack`. `TeachingDemos`, in turn, offers two functions: `faces`, another simplified version of that in `aplpack`, and `faces2`, which requires the input data to be provided as a matrix. For more details, read the documentation and examples provided in the help files for these packages. Examples of the use of faces and `faces2` for the exploration of the canine data are given in the supplementary resources of this book.

For stars, users just need to invoke the function `stars` from the core package `graphics`. In addition to the input matrix or data frame, further arguments can be given to get variants of the basic star plot, such as segment plots (circular sectors) and radar plots. See the R documentation for more details. Using the canine data, we provide on the book's website an example on the use of the function stars to produce a plot similar to Figure 4.3b, and a variant in which circular sectors are displayed for each variable.

A.5 Profiles of Variables

Profile plots can be generated with a combination of the high-level graphics functions `plot` and `barplot`, and low-level graphics functions (e.g., `axis` and `lines`). Low-level functions cannot be invoked if a high-level function has not been called earlier. This strategy assures that additional elements are placed in the plot without modifying the basic graph layout determined by the high-level function. As variables in Figure 3.5 are plotted in order of their average values, R function `order` is also needed. R scripts can be found in the book's website, producing the plots shown in Figures 3.5 and 3.6, with the latter plot using bars (via the `barplot` function) instead of lines.

References

Fox, J., and Weisberg, S. (2019). *An R Companion to Applied Regression*. 3rd Edn. Thousand Oaks, CA: SAGE. https://socialsciences.mcmaster.ca/jfox/Books/Companion/.

Grosjean, P. (2024). *SciViews-R*. Mons: UMONS. www.sciviews.org/SciViews-R.

Ligges, U., and Mächler, M. (2003). Scatterplot3d—an R package for visualizing multivariate data. *Journal of Statistical Software* 8(11): 1–20.

Murdoch, D., and Adler, D. (2024). *Rgl: 3D Visualization Using OpenGL*. R package version 1.3.1. https://CRAN.R-project.org/package=rgl.

Sarkar, D. (2008). *Lattice: Multivariate Data Visualization with R*. New York: Springer.

Schloerke, B., Cook, D., Larmarange, J., Briatte, F., Marbach, M., Thoen, E., Elberg, A., and Crowley, J. (2024). *GGally: Extension to 'ggplot2'*. R package version 2.2.1. https://CRAN.R-project.org/package=GGally.

Signorell, A. (2024). *DescTools: Tools for Descriptive Statistics*. R package version 0.99.54. https://CRAN.R-project.org/package=DescTools.

Snow, G. (2024). *TeachingDemos: Demonstrations for Teaching and Learning*. R package version 2.13. https://CRAN.R-project.org/package=TeachingDemos.

Wolf, H. (2019). *Aplpack: Another Plot Package (Version 190512)*. https://cran.r-project.org/package=aplpack.

4

Tests of Significance with Multivariate Data

4.1 Simultaneous Tests on Several Variables

When several variables are measured on the same sample units, it is possible to perform tests of significance on each one. For example, if the sample units are in two groups, then a difference between the means for the two groups can be tested separately for each variable. Unfortunately, there is a drawback to this simple approach because each test has some probability of leading to a wrong conclusion. In Section 4.4, we show that the probability of falsely finding at least one significant difference accumulates with the number of tests carried out.

One can adjust significance levels in such a case, but it may be preferable to conduct a single test that uses the information from all variables together. For example, one can test the hypothesis that the means of all variables are the same for two multivariate populations, with a significant result being taken as evidence that the means differ for at least one variable. This chapter studies these types of overall tests for the comparison of means and the comparison of variation for two or more samples.

4.2 Comparison of Mean Values for Two Samples: The Single-Variable Case

The first measurement in Table 1.1 on the body measurements of 49 female sparrows is total length. Is the mean of this variable the same for survivors and nonsurvivors of the storm that led to the birds being collected? Assuming the samples of 21 survivors and 28 nonsurvivors are effectively random samples from much larger populations of survivors and nonsurvivors, the question then is whether the observed difference in means is so large that it is unlikely to have occurred by chance if the population means are equal. A standard approach would be to carry out a t-test.

DOI: 10.1201/9781003453482-4

To generalize, given two random samples from different populations of a single variable X, let x_{i1} denote the values of X in the first sample, for $i = 1, 2, \ldots, n_1$, and x_{i2} denote the values in the second sample, for $i = 1, 2, \ldots, n_2$. The mean and variance for the jth sample are

$$\bar{x}_j = \frac{\sum_{i=1}^{n_j} x_{ij}}{n_j}$$

and

$$s_j^2 = \frac{\sum_{i=1}^{n_j} (x_{ij} - \bar{x}_j)^2}{n_j - 1} \tag{4.1}.$$

If we assume that X is normally distributed in both samples, with a common within-sample variance, a test to see whether the two-sample means are significantly different can be based on

$$t = \frac{\bar{x}_1 - \bar{x}_2}{s \cdot \sqrt{\dfrac{1}{n_1} + \dfrac{1}{n_2}}} \tag{4.2}$$

and seeing whether this is significantly different from zero by use of the t distribution with $n_1 + n_2 - 2$ degrees of freedom (df). Here,

$$s^2 = \frac{(n_1 - 1)s_1^2 + (n_2 - 1)s_2^2}{n_1 + n_2 - 2} \tag{4.3}$$

is the pooled estimate of variance from the two samples.

This test is fairly robust to the assumption of normality; if the population distributions of X are not too different from normal, it should be satisfactory, particularly for sample sizes of about 20 or more. The assumption of equal population variances is also not too crucial if the ratio of the true variances is within the limits 0.4–2.5. The test is particularly robust if the two sample sizes are equal, or nearly so.

If there are no concerns about nonnormality but the population variances may be unequal, then one possibility is to use a modified t-test. For example, Welch's (1951) test can be used

$$t = \frac{\bar{x}_1 - \bar{x}_2}{\sqrt{\dfrac{s_1^2}{n_1} + \dfrac{s_2^2}{n_2}}} \tag{4.4}.$$

The test proceeds as usual, using the t distribution with df equal to

$$df_w = \frac{(w_1 + w_2)^2}{\dfrac{w_1^2}{n_1 - 1} + \dfrac{w_2^2}{n_2 - 1}} \tag{4.5}$$

where $w_1 = s_1^2/n_1$, $w_2 = s_2^2/n_2$.

When there are concerns about both nonnormality and unequal variances, it has been shown by Manly and Francis (2002) that it may not be possible to reliably test for a difference in the population means. In particular, there may be too many significant results, providing evidence for a population mean difference when this does not really exist, irrespective of what test procedure is used. Manly and Francis provided a solution for this problem via a testing scheme that includes an assessment of whether two or more samples differ with respect to means or variances using randomization tests (Manly, 2009, section 4.6), and also an assessment of whether the randomization test for mean differences is reliable. See their paper for more details.

> **Author's comment:** It is common for many practitioners to follow the path of (1) check for equal variances. If they seem so, then (2a) uses the equal variances approach for the test; otherwise (2b) uses the Welch procedure. However, the biological "null" is that the two samples come from a single population of values, not from two. In that case, not only are the population means equal, but so are the population variances (since there is in fact only one population to begin with). The test is predicated on the means being equal (and then seeing how the evidence comes out). We believe that assuming equal variances (to begin with) is consistent with the biological null. On the other hand, when one estimates the size of a difference, one no longer assumes the samples come from a single population, and thus, there is no basis for equal variances either. For estimation, then, we use the Welch procedure. The two approaches do not often yield very different end results, and so we leave it to the practitioner to choose as they will.

4.3 Comparison of Mean Values for Two Samples: The Multivariate Case

For the female sparrow data that are shown in Table 1.1, a t-test can be employed for each of the five measurements shown in the table (total length, alar extent, length of beak and head, length of humerus, and length of keel of sternum). In that way, it is possible to decide which, if any, of these variables appear to have had different mean values for the populations of survivors and nonsurvivors. However, it may also be of some interest to know whether the five variables considered together suggest a difference between survivors and nonsurvivors. Does the total evidence point to mean differences between the populations of surviving and nonsurviving sparrows?

A multivariate test can answer this question. One possibility is Hotelling's T^2-test, which is a generalization of the t-test of Equation 4.2 or, to be more precise, the square of this t-statistic.

In a general case, there will be p variables $X_1, X_2, \ldots, X_p$, and two samples with sizes n_1 and n_2. There are then two sample mean vectors, $\bar{x}_1$ and $\bar{x}_2$, and two sample covariance matrices, $\mathbf{C}_1$ and $\mathbf{C}_2$, with these being calculated as explained in Section 2.7.

Assuming that the population covariance matrices are the same for both populations, a pooled estimate of this matrix is

$$\mathbf{C} = \frac{(n_1 - 1)\mathbf{C}_1 + (n_2 - 1)\mathbf{C}_2}{n_1 + n_2 - 2} \tag{4.6.}$$

Hotelling's T^2-statistic is then defined as

$$T^2 = \frac{n_1 \cdot n_2 (\bar{\mathbf{x}}_1 - \bar{\mathbf{x}}_2)' \mathbf{C}^{-1}(\bar{\mathbf{x}}_1 - \bar{\mathbf{x}}_2)}{n_1 + n_2} \tag{4.7.}$$

A significantly large value for this statistic is evidence that the two population mean vectors are different. The significance, or lack of significance, of T^2 is most simply determined by using the fact that if the null hypothesis of equal population mean vectors is true, then the transformed statistic

$$F = \frac{(n_1 + n_2 - p - 1)T^2}{(n_1 + n_2 - 2)p} \tag{4.8}$$

follows an F distribution with p and $(n_1 + n_2 - p - 1)$ df.

The T^2-statistic is a quadratic form, as defined in Section 2.5. It can, therefore, be written as the double sum

$$T^2 = \frac{n_1 n_2}{n_1 + n_2} \sum_{i=1}^{p} \sum_{k=1}^{p} (\bar{x}_{1i} - \bar{x}_{2i}) c_{(ik)} (\bar{x}_{1k} - \bar{x}_{2k}) \tag{4.9.}$$

which may be simpler to compute. Here, $\bar{x}_{ji}$ is the mean of variable X_i in the jth sample, and $c_{(ik)}$ is the element in the ith row and kth column of the inverse matrix $\mathbf{C}^{-1}$.

The two samples being compared using the T²-statistic are assumed to come from multivariate normal distributions with equal covariance matrices. Some deviation from multivariate normality is probably not serious.

A moderate difference between population covariance matrices is also not too important, particularly with equal or nearly equal sample sizes. If the two population covariance matrices are very different, and sample sizes are very different as well, then a modified test can be used (Yao, 1965), but this still relies on the assumption of multivariate normality.

4.3.1 Example 4.1: Testing Mean Values for Bumpus' Female Sparrows

For the female sparrow data from Table 1.1, one can ask whether there are any differences between survivors and nonsurvivors with respect to the mean values of five morphological characters.

We will use X_1, the total length to exemplify the single sample method. The mean of this variable for the 21 survivors is $\bar{x}_1 = 157.38$, while the mean for the 28 nonsurvivors is $\bar{x}_2 = 158.43$. The corresponding sample variances are $s_1^2 = 11.05$ and $s_2^2 = 15.07$. The pooled variance from Equation 4.3 is therefore

$$s^2 = (20 \times 11.05 + 27 \times 15.07)/47 = 13.36$$

and the t-statistic of Equation 4.2 is

$$t = (157.38 - 158.43)/\sqrt{[13.36(1/21 + 1/28)]} = -0.99$$

with $n_1 + n_2 - 2 = 47\,\mathrm{df}$. There is no evidence of a difference in population means of total length between survivors and nonsurvivors.

Table 4.1 summarizes the results of tests on all five of the variables taken individually. In no case is there any evidence of a population mean difference between survivors and nonsurvivors.

TABLE 4.1

Comparison of Mean Values for Survivors and Nonsurvivors for Bumpus' Female Sparrows with Variables Taken One at a Time

Variable	Survivors		Nonsurvivors		t (47 df)	p-value[a]
	$\bar{x}_1$	s_1^2	$\bar{x}_2$	s_2^2		
Total length	157.38	11.05	158.43	15.07	−0.99	0.326
Alar extent	241.00	17.50	241.57	32.55	−0.39	0.700
Length of beak and head	31.43	0.53	31.48	0.73	−0.20	0.846
Length of humerus	18.50	0.18	18.45	0.43	0.33	0.746
Length of keel of sternum	20.81	0.57	20.84	1.32	−0.10	0.919

[a] *Probability of obtaining a t-value as far from zero as the observed value if the null hypothesis of no population mean difference is true.*

For tests on all five variables considered together, we would recommend using statistical software, but for sake of illustration we will show you the calculations. For the sample of 21 survivors, the mean vector and covariance matrix are

$$
\bar{x}_1 = \begin{bmatrix} 157.381 \\ 241.000 \\ 31.433 \\ 18.500 \\ 20.810 \end{bmatrix} \text{ and } C_1 = \begin{bmatrix} 11.048 & 9.100 & 1.557 & 0.870 & 1.286 \\ 9.100 & 17.500 & 1.910 & 1.310 & 0.880 \\ 1.557 & 1.910 & 0.531 & 0.189 & 0.240 \\ 0.870 & 1.310 & 0.189 & 0.176 & 0.133 \\ 1.286 & 0.880 & 0.240 & 0.133 & 0.575 \end{bmatrix}.
$$

For the sample of 28 nonsurvivors, the results are

$$
\bar{x}_2 = \begin{bmatrix} 158.429 \\ 241.571 \\ 31.479 \\ 18.446 \\ 20.839 \end{bmatrix} \text{ and } C_2 = \begin{bmatrix} 15.069 & 17.190 & 2.243 & 1.746 & 2.931 \\ 17.190 & 32.550 & 3.398 & 2.950 & 4.066 \\ 2.243 & 3.398 & 0.728 & 0.470 & 0.559 \\ 1.746 & 2.950 & 0.470 & 0.434 & 0.506 \\ 2.931 & 4.066 & 0.559 & 0.506 & 1.321 \end{bmatrix}.
$$

The pooled sample covariance is then

$$
C = (20 \cdot C_1 + 27 \cdot C_2)/47 = \begin{bmatrix} 13.358 & 13.748 & 1.951 & 1.373 & 2.231 \\ 13.748 & 26.146 & 2.765 & 2.252 & 2.710 \\ 1.951 & 2.765 & 0.645 & 0.350 & 0.423 \\ 1.373 & 2.252 & 0.350 & 0.324 & 0.347 \\ 2.231 & 2.710 & 0.423 & 0.347 & 1.004 \end{bmatrix}
$$

where, for example, the element in the second row and third column is

$$
(20 \times 1.910 + 27 \times 3.398)/47 = 2.765.
$$

The inverse of the matrix C is found to be

$$
C^{-1} = \begin{bmatrix} 0.2061 & -0.0694 & -0.2395 & 0.0785 & -0.1969 \\ -0.0694 & 0.1234 & -0.0376 & -0.5517 & 0.0277 \\ -0.2395 & -0.0376 & 4.2219 & -3.2624 & -0.0181 \\ 0.0785 & -0.5517 & -3.2624 & 11.4610 & -1.2720 \\ -0.1969 & 0.0277 & -0.0181 & -1.2720 & 1.8068 \end{bmatrix}.
$$

This can be verified by evaluating the product $\mathbf{C} \cdot \mathbf{C}^{-1}$ and seeing that this is a unit matrix, apart from rounding errors.

Substituting the elements of $\mathbf{C}^{-1}$ and other values into Equation 4.7 produces

$$
\begin{aligned}
T^2 = [(21 \times 28)/(21+28)][&(157.381-158.429) \times 0.2061 \times (157.381-158.429) \\
&+(241.000-241.571) \times (-0.0694) \times (241.000-241.571) \\
&+(31.433-31.479) \times (-0.2395) \times (31.433-31.479)+...+ \\
&(20.810-20.839) \times (1.8068) \times (20.810-20.839)] = 2.824
\end{aligned}
$$

Using Equation 4.8, this converts to an F-statistic of

$$
F = (21+28-5-1) \times 2.824/[(21+28-2) \times 5] = 0.517
$$

with 5 and 43 df. Clearly, this is not significantly large, because a significant F-value must exceed unity. Hence, there is no evidence of a difference in population means for survivors and non-survivors, taking all five variables together.

4.4 Multivariate versus Univariate Tests

In the last example, there were no significant results either for the variables considered individually or for the overall multivariate test. However, it is quite possible to have insignificant univariate tests but a significant multivariate test. This can occur because of the accumulation of the evidence from the individual variables in the overall test. Conversely, an insignificant multivariate test can occur when some univariate tests are significant, because the evidence of a difference provided by the significant variables is swamped by the evidence of no difference provided by the other variables.

One important aspect of the use of a multivariate test as distinct from a series of univariate tests concerns the control of type one error rates (a so-called false significance). With a univariate test at the 5% level, there is a 0.95 probability of a nonsignificant result when the population means are the same. Hence, if p independent tests are carried out under these conditions, then the probability of getting no significant results is 0.95^p. The probability of at least one significant result is therefore $1-0.95^p$, which may be unacceptably large. For example, if $p = 5$, then the probability of at least one significant result by chance alone is $1-0.95^5 = 0.23$. With multivariate data, variables are usually not independent, so $1-0.95^p$ does not quite give the correct probability. However, the principle still applies that more tests lead to a higher probability of obtaining at least one significant result by chance.

On the other hand, a multivariate test such as Hotelling's T^2 test using the 5% level of significance gives a 0.05 probability of a type one error, irrespective of the number of variables involved, providing that the assumptions of the test hold. This is a distinct advantage over a series of univariate tests, particularly when the number of variables is large.

There are ways of adjusting significance levels to control the overall probability of a type one error when several univariate tests are carried out. The simplest approach involves using a Bonferroni adjustment. If p tests are carried out using the significance level (α/p), then the probability of obtaining any significant results by chance is less than α.

Some people are not inclined to use a Bonferroni correction to significance levels, because the significance levels applied to the individual tests become so extreme if p is large. For example, with $p = 10$ and an overall 0.05 level of significance, a univariate test result is only declared significant if it is significant at the 0.005 level. This has led to the development of some slightly less conservative variations on the Bonferroni correction, as discussed by Manly (2009, section 4.9).

It can certainly be argued that the use of a single multivariate test provides a better procedure in many cases than making a large number of univariate tests. A multivariate test also has the added advantage of taking proper account of the correlation between variables.

4.5 Comparison of Variation for Two Samples: The Single-Variable Case

With a single variable, the best-known method for comparing the variation in two samples is the F-test. If s_j^2 is the variance in the jth sample, calculated as shown in Equation 4.1, then the ratio s_1^2/s_2^2 is compared with percentage points of the F distribution with $(n_1 - 1)$ and $(n_2 - 1)$ df. A value of the ratio that is significantly different from one is then evidence that the samples are from two populations with different variances. Unfortunately, the F-test is known to be rather sensitive to the assumption of normality. A significant result may well be due to the fact that a variable is not normally distributed rather than to unequal variances. For this reason, it is sometimes argued that the F-test should never be used to compare variances.

A robust alternative to the F-test is Levene's (1960) test. The idea here is to transform the original data in each sample into absolute deviations from the sample mean or the sample median and then test for a significant difference between the mean deviations in the two samples, using a t-test. Although absolute deviations from the sample means are sometimes used, a more robust test is likely to be obtained by using absolute deviations from the sample medians (Schultz, 1983). The procedure using medians is illustrated in Example 4.3.

4.6 Comparison of Variation for Two Samples: The Multivariate Case

Many computer packages use Box's M-test to compare the variation in two or more multivariate samples. Because this applies for two or more samples, it is described in Section 4.8. This test is known to be rather sensitive to the assumption that the samples are from multivariate normal distributions. A significant result could arise due to nonnormality rather than to unequal population covariance matrices.

An alternative procedure that should be more robust can be constructed using the principle behind Levene's test. Thus, the data values can be transformed into absolute deviations from sample means or medians. If the median is chosen, the *absolute deviation from the sample median* is

$$ADM_{ijk} = \left| x_{ijk} - M_{jk} \right| \tag{4.10}$$

where

x_{ijk} is the value of variable X_k for the ith individual in sample j;

M_{jk} is the median of the same variable in the sample.

The unequal variation question becomes a test of multivariate means in the *ADM* values,

$$\overline{ADM}_{jk} = \frac{\sum_{i=1}^{n_j} ADM_{ijk}}{n_j} \tag{4.11}.$$

which can be done using a T^2-test

Van Valen (1978) suggested another possibility for the comparison of the variation in two multivariate samples. This involves calculating

$$d_{ij} = \sqrt{\sum_{k=1}^{p} (x_{ijk} - \bar{x}_{jk})^2} \tag{4.12}$$

where

x_{ijk} is the value of variable X_k for the ith individual in sample j;

$\bar{x}_{jk}$ is the mean of the same variable in the sample.

The sample means of the d_{ij} values are compared with a t-test. If one sample is more variable than another, then the mean d_{ij} values will tend to be higher in that sample.

To ensure that all variables are given equal weight, each variable should be standardized so that the mean is zero and variance is one for all samples combined before the calculation of the d_{ij} values. For a more robust test, it may be better to use sample medians in place of the sample means in Equation 4.12. Then, the formula for d_{ij} values is

$$d_{ij} = \sqrt{\sum_{k=1}^{p}(x_{ijk} - M_{jk})^2} \tag{4.13}$$

where M_{jk} is the median for variable X_k in the jth sample. Then the signed deviations from medians $(x_{ijk} - M_{jk})$ are used in van Valen's test. As $(x_{ijk} - M_{jk})^2 = |x_{ijk} - M_{jk}|^2$, the d_{ij} values given by in Equation 4.13 can also be computed using the *ADM*s.

The T^2-test for absolute deviation from medians and Van Valen's test for deviations from medians are illustrated in the example that follows. One point to note about the use of the test statistics 4.12 and 4.13 is that they are based on an implicit assumption that if the two samples being tested differ, then one sample will be more variable than the other for all variables. A significant result cannot be expected in a case where, for example, X_1 and X_2 are more variable in sample 1, but X_3 and X_4 are more variable in sample 2. The effect of the differing variances would then tend to cancel out in the calculation of d_{ij}. Thus, Van Valen's test is not appropriate for situations where changes in the level of variation are not expected to be consistent for all variables.

4.6.1 Example 4.3: Testing Variation for Female Sparrows

With Bumpus' female sparrow data shown in Table 1.1, one interesting question concerns whether the nonsurvivors are more variable than the survivors. This is what would be expected if stabilizing selection took place.

To examine this question, the individual variables can be considered one at a time, starting with X_1, the total length. For Levene's test, the original data values are transformed into deviations from sample medians. The median for survivors is 157 mm, and the absolute deviations from this median for the 21 birds in the sample then have a mean of $\bar{x}_1 = 2.571$ and a variance of $s_1^2 = 4.257$. The median for nonsurvivors is 159 mm, and the absolute deviations from this median for the 28 birds in the sample have a mean of $\bar{x}_2 = 3.286$ with a variance of $s_2^2 = 4.212$. The pooled variance from Equation 4.3 is 4.231, and the t-statistic of Equation 4.2 is

$$t = (2.578 - 3.29)/[4.231(1/21 + 1/28)]^{1/2} = -1.20$$

with 47 df.

Because nonsurvivors would be more variable than survivors if stabilizing selection occurred, it is a one-sided test that is required here, with low values of t providing evidence of selection. The observed value of t is not significantly low in the present instance. The t-values for the other variables are as follows: the alar extent, $t = -1.18$; the length of the beak and head, $t = -0.81$; the length of the humerus, $t = -1.91$; and the length of keel of the sternum, $t = -1.41$. Only for the length of the humerus is the result significantly low at the 5% level.

To illustrate the comparison of variation for two samples in the multivariate case, Hotelling's T^2-test is considered first, based on the idea that two samples can be compared in terms of the means of absolute deviations from medians $(\overline{ADM})$, similar to the principle behind Levene's test. Table 4.2 shows the absolute deviations from sample medians for the sparrow data. For example, the median for total length is 157.0. The measurement on this variable for the first surviving female sparrow was 156.0, and the value given in Table 4.2 is 1.0, obtained as $|156.0–157.0| = 1.0$

TABLE 4.2

Medians and Absolute Deviations from Sample Medians (Equation 4.10) for the Data on Size of the Surviving (S) and Nonsurviving (NS) Female Sparrows as Shown in Table 1.1; $\overline{ADM}$= Mean Absolute Deviation from the Sample Median, Given by Equation 4.11

	Total length	Alar extent	Beak and head	Length of humerus	Keel of sternum
Medians for S	157.0	240.0	31.4	18.5	20.6
Medians for NS	159.0	242.0	31.5	18.5	20.7
	Absolute deviations from sample medians				
Survival	**Total length**	**Alar extent**	**Beak and head**	**Length of humerus**	**Keel of sternum**
S	1.0	5.0	0.2	0.0	0.1
S	3.0	0.0	1.0	0.6	1.0
S	4.0	0.0	0.4	0.1	0.0
S	4.0	4.0	0.5	0.8	0.4
S	2.0	3.0	0.1	0.1	0.3
S	6.0	7.0	0.6	0.5	0.3
S	0.0	2.0	0.5	0.1	0.4
S	2.0	1.0	1.4	0.1	0.6
S	7.0	8.0	1.3	0.6	0.5
S	1.0	2.0	0.4	0.3	1.4
S	1.0	0.0	0.1	0.1	1.4

(Continued)

Multivariate Statistical Methods

TABLE 4.2 (Continued)

	Total length	Alar extent	Beak and head	Length of humerus	Keel of sternum
S	3.0	4.0	0.3	0.1	0.1
S	4.0	6.0	0.9	0.8	1.2
S	0.0	5.0	0.6	0.6	0.6
S	0.0	5.0	0.1	0.4	0.8
S	1.0	3.0	0.5	0.5	0.3
S	1.0	4.0	0.0	0.0	1.0
S	4.0	2.0	0.9	0.3	0.3
S	2.0	4.0	1.1	0.0	0.5
S	6.0	6.0	1.1	0.1	1.3
S	2.0	4.0	0.1	0.5	0.9
$\overline{ADM}$(S)	2.57	3.57	0.58	0.31	0.64
NS	4.0	2.0	0.1	0.5	0.0
NS	3.0	2.0	0.0	0.3	0.1
NS	1.0	0.0	1.1	0.3	1.0
NS	7.0	10.0	1.2	1.3	0.9
NS	1.0	8.0	0.2	0.3	1.8
NS	4.0	5.0	0.5	0.0	0.7
NS	2.0	3.0	0.7	1.0	0.7
NS	6.0	3.0	1.6	1.3	2.0
NS	6.0	11.0	1.4	1.2	0.9
NS	3.0	3.0	1.2	0.5	2.4
NS	3.0	1.0	0.1	0.3	0.6
NS	0.0	3.0	0.3	0.0	1.0
NS	0.0	5.0	0.6	0.4	1.7
NS	4.0	1.0	0.6	0.0	0.6
NS	3.0	10.0	0.4	0.6	1.5
NS	7.0	12.0	1.1	1.2	2.1
NS	0.0	0.0	0.7	0.3	0.2
NS	4.0	4.0	0.3	0.6	1.4
NS	4.0	7.0	1.9	1.0	2.1
NS	4.0	0.0	0.5	0.4	0.0
NS	3.0	5.0	0.2	0.3	0.4
NS	0.0	4.0	0.0	0.1	0.4

(*Continued*)

TABLE 4.2   (Continued)

	Total length	Alar extent	Beak and head	Length of humerus	Keel of sternum
NS	2.0	3.0	0.6	0.6	0.1
NS	4.0	7.0	0.8	0.8	1.1
NS	3.0	5.0	0.4	0.6	0.3
NS	6.0	5.0	0.9	0.1	0.3
NS	3.0	3.0	1.0	0.0	0.4
NS	5.0	6.0	0.8	0.3	0.2
$\overline{ADM}$ (NS)	3.29	4.57	0.69	0.51	0.89

Comparing the sample $\overline{ADM}$ vectors for the five variables using Hotelling's T^2-test gives a test statistic of $T^2 = 4.748$, corresponding to an F-statistic of 0.869 with 5 and 43 df using Equation 4.8. There is, therefore, no evidence of a significant difference between the samples from this test, because the F-value is less than one (0.869; p = 0.51).

Finally, consider Van Valen's test. Table 4.3 shows the deviations from sample medians for the data after they have been standardized. For example, the first value given for variable Total Length, for survivors, is −0.274. This was obtained as follows. First, the original data were coded to have a zero mean and a unit variance for all 49 birds. This transformed the total length for the first survivor to $(156 − 157.980)/3.654 = −0.542$. The median transformed length for survivors was then −0.268. Hence, the deviation from the sample median for the first survivor is $−0.542 − (−0.268) = −0.274$, as shown in Table 4.3. The d-values from Equation 4.13, that is, the sums on the right-hand side of the equation for individuals within samples, are shown in the last column of Table 4.3. The mean for survivors is 1.742, with variance 0.402. The mean for nonsurvivors is 2.242, with variance 1.110. The t-value from Equation 4.2 is then −1.92, which is significantly low at the 5% level (p = 0.03 on a one-sided test). Hence, this test indicates more variation for nonsurvivors than for survivors.

An explanation for the significant result with Van Valen's test, but no significant result with Levene's test based on Hotelling's T^2, is not hard to find. As noted earlier, Levene's test is not directional and does not take into account the expectation that the survivors will, if anything, be less variable than the nonsurvivors. On the other hand, Van Valen's test is specifically for less variation in Sample 1 than in Sample 2, for all variables. In the present case, all the variables show less variation in Sample 1 than in Sample 2. Van Valen's test has emphasized this fact, but Levene's test has not.

TABLE 4.3

Standardized Values and Deviations from Sample Medians of the Standardized Values for the Data on Size of the Surviving (S) and Nonsurviving (NS) Female Sparrows as Shown in Table 1.1

Survival	Standardized data					Deviations from sample medians					
	Total length	Alar extent	Beak and head	Length of humerus	Keel of sternum	Total length	Alar extent	Beak and head	Length of humerus	Keel of sternum	d
S	-0.542	0.725	0.177	0.054	-0.329	-0.274	0.987	0.252	0.000	-0.101	1.059
S	-1.089	-0.262	-1.333	-1.009	-1.237	-0.821	0.000	-1.258	-1.063	-1.009	2.099
S	-1.363	-0.262	-0.578	-0.123	-0.229	-1.095	0.000	-0.503	-0.177	0.000	1.218
S	-1.363	-1.051	-0.704	-1.363	-0.632	-1.095	-0.789	-0.629	-1.418	-0.403	2.095
S	-0.815	0.330	0.051	0.231	-0.531	-0.547	0.592	0.126	0.177	-0.303	0.888
S	1.374	1.120	0.680	0.940	0.074	1.642	1.381	0.755	0.886	0.303	2.460
S	-0.268	-0.656	-0.704	-0.123	-0.632	0.000	-0.395	-0.629	-0.177	-0.403	0.864
S	-0.815	-0.459	1.687	0.231	0.377	-0.547	-0.197	1.762	0.177	0.605	1.959
S	1.647	1.317	1.561	1.118	0.276	1.916	1.579	1.636	1.063	0.504	3.197
S	0.006	-0.656	-0.578	0.586	1.184	0.274	-0.395	-0.503	0.532	1.412	1.662
S	0.006	-0.262	-0.200	0.231	1.184	0.274	0.000	-0.126	0.177	1.412	1.455
S	0.553	0.528	-0.452	0.231	-0.329	0.821	0.789	-0.377	0.177	-0.101	1.217
S	0.827	0.922	1.058	1.472	0.982	1.095	1.184	1.132	1.418	1.210	2.712
S	-0.268	0.725	0.680	1.118	-0.834	0.000	0.987	0.755	1.063	-0.605	1.744
S	-0.268	-1.248	0.051	-0.655	-1.035	0.000	-0.987	0.126	-0.709	-0.807	1.464
S	-0.542	-0.854	-0.704	-0.832	-0.531	-0.274	-0.592	-0.629	-0.886	-0.303	1.303
S	0.006	0.528	-0.074	0.054	0.780	0.274	0.789	0.000	0.000	1.009	1.310

(*Continued*)

TABLE 4.3 (Continued)

Survival	Standardized data					Deviations from sample medians					
	Total length	Alar extent	Beak and head	Length of humerus	Keel of sternum	Total length	Alar extent	Beak and head	Length of humerus	Keel of sternum	d
S	−1.363	−0.656	−1.207	−0.477	0.074	−1.095	−0.395	−1.132	−0.532	0.303	1.735
S	−0.815	−1.051	−1.459	0.054	−0.733	−0.547	−0.789	−1.384	0.000	−0.504	1.759
S	1.374	0.922	1.310	0.231	1.083	1.642	1.184	1.384	0.177	1.311	2.786
S	0.279	−1.051	0.051	−0.832	0.679	0.547	−0.789	0.126	−0.886	0.908	1.596
Median	−0.268	−0.262	−0.074	0.054	−0.229					Mean	1.742
										Var	0.402
NS	−0.815	−0.262	−0.074	−0.832	−0.128	−1.095	−0.395	−0.126	−0.886	0.000	1.468
NS	−0.542	−0.262	0.051	−0.477	−0.229	−0.821	−0.395	0.000	−0.532	−0.101	1.059
NS	0.553	0.133	1.435	0.586	0.881	0.274	0.000	1.384	0.532	1.009	1.814
NS	−1.636	−1.840	−1.459	−2.250	−1.035	−1.916	−1.973	−1.510	−2.304	−0.908	3.997
NS	0.553	1.711	0.303	0.586	1.688	0.274	1.579	0.252	0.532	1.816	2.492
NS	−0.815	−0.854	−0.578	0.054	−0.834	−1.095	−0.987	−0.629	0.000	−0.706	1.751
NS	−0.268	0.725	0.932	1.826	0.578	−0.547	0.592	0.881	1.772	0.706	2.251
NS	1.921	0.725	2.065	2.358	1.890	1.642	0.592	2.013	2.304	2.017	4.059
NS	−1.363	−2.038	−1.710	−2.072	−1.035	−1.642	−2.171	−1.762	−2.127	−0.908	3.982
NS	1.100	−0.459	−1.459	−0.832	2.293	0.821	−0.592	−1.510	−0.886	2.421	3.154
NS	1.100	0.330	0.177	0.586	0.478	0.821	0.197	0.126	0.532	0.605	1.174
NS	0.279	0.725	0.429	0.054	0.881	0.000	0.592	0.377	0.000	1.009	1.229
NS	0.279	1.120	−0.704	−0.655	−1.842	0.000	0.987	−0.755	−0.709	−1.715	2.233

(Continued)

TABLE 4.3 (Continued)

Survival	Standardized data					Deviations from sample medians					
	Total length	Alar extent	Beak and head	Length of humerus	Keel of sternum	Total length	Alar extent	Beak and head	Length of humerus	Keel of sternum	d
NS	-0.815	0.330	-0.704	0.054	0.478	-1.095	0.197	-0.755	0.000	0.605	1.474
NS	1.100	2.106	0.555	1.118	1.385	0.821	1.973	0.503	1.063	1.513	2.871
NS	-1.636	-2.235	-1.333	-2.072	-2.246	-1.916	-2.368	-1.384	-2.127	-2.118	4.495
NS	0.279	0.133	-0.829	-0.477	-0.329	0.000	0.000	-0.881	-0.532	-0.202	1.048
NS	-0.815	-0.656	-0.326	-1.009	-1.540	-1.095	-0.789	-0.377	-1.063	-1.412	2.256
NS	1.374	1.514	2.442	1.826	1.991	1.095	1.381	2.391	1.772	2.118	4.056
NS	1.374	0.133	-0.578	-0.655	-0.128	1.095	0.000	-0.629	-0.709	0.000	1.448
NS	-0.542	-0.854	0.303	-0.477	-0.531	-0.821	-0.987	0.252	-0.532	-0.403	1.468
NS	0.279	-0.656	0.051	-0.123	-0.531	0.000	-0.789	0.000	-0.177	-0.403	0.904
NS	0.827	0.725	0.806	1.118	-0.027	0.547	0.592	0.755	1.063	0.101	1.536
NS	-0.815	-1.248	-0.955	-1.363	-1.237	-1.095	-1.381	-1.007	-1.418	-1.110	2.713
NS	1.100	1.120	0.555	1.118	-0.430	0.821	0.987	0.503	1.063	-0.303	1.767
NS	-1.363	-0.854	-1.081	0.231	-0.430	-1.642	-0.987	-1.132	0.177	-0.303	2.253
NS	1.100	0.725	1.310	0.054	0.276	0.821	0.592	1.258	0.000	0.403	1.664
NS	1.647	1.317	1.058	0.586	0.074	1.368	1.184	1.007	0.532	0.202	2.147
Median	0.279	0.133	0.051	0.054	-0.128					Mean	2.242
										Var	1.110

Note: The standardized values have a mean of zero and a standard deviation of one for all of the data combined, for all five variables. These values are used for the Van Valen (1978) test to compare the variation in the two samples of survivors and nonsurvivors based on deviations from sample medians and Equation 4.13. The Van Valen test compares the two sample means of d-values using a t-test. This shows that there is significantly more variation in size for nonsurvivors than for survivors ($t = -1.92$ with 47 df, p-value = 0.030 on a one-sided test).

4.7 Comparison of Means for Several Samples

When there is a single variable and samples from several populations to be compared, the generalization of the t-test is the F-test from a one-factor analysis of variance (ANOVA). The calculations are as shown in Table 4.4.

When there are several variables and several samples, the situation is complicated by the fact that there are four alternative statistics that are commonly used to test the hypothesis that all the samples came from populations with the same mean vector. This is the simplest case of what is sometimes called *multivariate analysis of variance* (MANOVA) or, more precisely, *one-factor MANOVA*, to emphasize that only one factor is involved.

The first test to be considered uses *Wilks' lambda statistic*

$$\Lambda = |\mathbf{W}|/|\mathbf{T}| \qquad (4.14)$$

where

$|\mathbf{W}|$ is the determinant of the *within-sample sum of squares and cross products matrix;*

$|\mathbf{T}|$ is the determinant of the *total sum of squares and cross products matrix.*

Essentially, this compares the variation within the samples with the variation both within and between the samples. Here, the matrices $\mathbf{T}$ and $\mathbf{W}$ require some further explanation. Let x_{ijk} denote the value of variable X_k for the ith individual in the jth sample, let $\bar{x}_{jk}$ denote the mean of X_k in the same sample, and let $\bar{x}_k$ denote the overall mean of X_k for all the data taken together. In

TABLE 4.4

One-Factor Analysis of Variance (ANOVA) for Comparing the Mean Values of Samples from m Populations, with a Single Variable

Source of variation	Sum of squares	df	Mean square	F-ratio
Between samples	$B = T - W$	$m - 1$	$M_B = B/(m-1)$	$F = M_B / M_W$
Within samples	$W = \sum_{j=1}^{m}\sum_{i=1}^{n_j}(x_{ij} - \bar{x}_j)^2$	$n - m$	$M_W = W/(n-m)$	
Total	$T = \sum_{j=1}^{m}\sum_{i=1}^{n_j}(x_{ij} - \bar{x})^2$	$n - 1$		

Note: n_j = size of the jth sample; $n = n_1 + n_2 + \cdots + n_m$ = total number of observations; x_{ij} = ith observation in the jth sample; $\bar{x}_j$ = mean of the jth sample; $\bar{x}$ = mean of all observations.

addition, assume that there are m samples, with the jth of size n_j. Then, the element in row r and column c of $\mathbf{T}$ is

$$t_{rc} = \sum_{j=1}^{m}\sum_{i=1}^{n_j}(x_{ijr} - \bar{x}_r)(x_{ijc} - \bar{x}_c) \tag{4.15}$$

and the element in row r and column c of $\mathbf{W}$ is

$$w_{rc} = \sum_{j=1}^{m}\sum_{i=1}^{n_j}(x_{ijr} - \bar{x}_{jr})(x_{ijc} - \bar{x}_{jc}). \tag{4.16}$$

Determinants are introduced in Section 2.4. Here, it is enough to understand that they are scalar quantities, and that special computer algorithms are needed to calculate them unless the matrices involved are of size 2×2, or possibly 3×3.

If Λ is small, this indicates that the variation within the samples is low in comparison with the total variation. This, then, provides evidence that the samples do not come from populations with the same mean vector. An approximate test for whether the within-sample variation is significantly low in this respect is described in Table 4.5. Tables of exact critical values are also available.

Let $\lambda_1 \geq \lambda_2 \geq ... \geq \lambda_p \geq 0$ be the eigenvalues of $\mathbf{W}^{-1}\mathbf{B}$, where $\mathbf{B} = \mathbf{T} - \mathbf{W}$ is called the *between-sample matrix of sums of squares and cross products*, because the typical entry is the difference between a total sum of squares or cross product minus the corresponding term within samples. Then, Wilks' lambda can also be expressed as

$$\Lambda = \prod_{i=1}^{p}1/(1+\lambda_i) \tag{4.17},$$

which is the form that is sometimes used to represent it.

A second statistic is the largest eigenvalue λ_1 of the matrix $\mathbf{W}^{-1}\mathbf{B}$, which leads to what is called *Roy's largest root test* (eigenvalues are also called *latent roots*). The basis for using this statistic is the result that if the linear combination of the variables $X_1 - X_p$ that maximizes the ratio of the between-sample sum of squares to the within-sample sum of squares is found, then this maximum ratio equals λ_1. This maximum eigenvalue can test whether the between-sample variation is significantly large (evidence that the samples do not come from populations with the same mean vector). This approach is related to discriminant function analysis (Chapter 8). Note that what some computer programs call Roy's largest root statistic is $\lambda_1/(1-\lambda_1)$ rather than λ_1 itself. If in doubt, check the program documentation.

TABLE 4.5

Test Statistics Used to Compare Sample Mean Vectors (One-Factor MANOVA) with Approximate F tests for Evidence That the Population Values Are Not Constant

Test	Statistic	F	df₁	df₂	Comment
Wilks' lambda Λ	Λ	$[(1-\Lambda^{1/t})/\Lambda^{1/t}](df_2/df_1)$	$p(m-1)$	$w\cdot t - df_1/2+1$	$w=n-1-(p+m)/2$ $t=\{(df_1^2-4)/[p^2+(m-1)^2-5]\}^{1/2}$ If $df_1=2$, set $t=1$.
Roy's largest root λ_1	λ_1	$(df_2/df_1)\lambda_1$	d	$n-m-d-1$	The significance level obtained is a lower bound. $d=\max(p,m-1)$
Pillai's trace	$V=\sum_{i=1}^{p}\lambda_i/(1+\lambda_i)$	$(n-m-p+s)V/[d(s-V)]$	$s\cdot d$	$s(n-m-p+s)$	$s=\min(p,m-1)=$ number of positive eigenvalues $d=\max(p,m-1)$
Lawley-Hotelling trace	$U=\sum_{i=1}^{p}\lambda_i$	$df_2\cdot U/(s\cdot df_1)$	$s(2A+s+1)$	$2(s\cdot B+1)$	s is as for Pillai's trace $A=(|m-p-1|-1)/2$ $B=(n-m-p-1)/2$

Note: It is assumed that there are p variables in m samples, with the jth of size n_j, and a total sample size of $n = \Sigma n_j$. These are approximations for general p and m. Exact or better approximations are available for some special cases, and other approximations are also available. In all cases, the test statistic is transformed to the stated F value, and this is tested to see whether it is significantly large in comparison with the F distribution with df_1 and df_2 degrees of freedom. Chi-squared distribution approximations are also in common use, and tables of critical values are available (Kres, 1983).

To assess whether λ_1 is significantly large, a p-value can be calculated numerically, or an F distribution can be used to find a lower bound to the significance level. That is the true significance level is greater than the p-value generated by the F test. Computer package users should be aware of which of these alternatives is used. If the F distribution is used, then the value of λ_1 may not actually be significantly large at the chosen significance level. The F-value used is described in Table 4.5.

Pillai's trace statistic is a third way to test whether the samples come from populations with the same mean vectors. This can be written in terms of the eigenvalues $\lambda_1 - \lambda_p$ as

$$V = \sum_{i=1}^{p} \lambda_i / (1 + \lambda_i) \tag{4.18}.$$

Again, large values of this statistic provide evidence that the samples being considered come from populations with different mean vectors. An approximation for the significance level is provided in Table 4.5.

Finally, the fourth statistic often used to test the null hypothesis of equal population mean vectors is the *Lawley–Hotelling trace*

$$U = \sum_{i=1}^{p} \lambda_i \tag{4.19}$$

which is just the sum of the eigenvalues of the matrix $\mathbf{W}^{-1}\mathbf{B}$. Yet again, large values provide evidence against the null hypothesis, with an approximate F-test provided in Table 4.5.

Generally, the four tests just described can be expected to give similar significance levels so that there is no real need to choose between them. They all involve the assumption that the distribution of the p variables is multivariate normal, with the same within-sample covariance matrix for all the m populations from which the samples are drawn. They are all also considered to be fairly robust if the sample sizes are equal or nearly so for the m samples. If there are questions about either the multivariate normality or the equality of covariance matrices, then simulation studies suggest that Pillai's trace statistic may be more robust than the other three statistics (Seber, 2004, p. 442).

4.8 Comparison of Variation for Several Samples

Box's M-test is the best known for comparing the variation in several samples. This test has already been mentioned for the two-sample situations with several variables to be compared, and it can be used with one or several variables, with two or more samples.

For m samples, the M statistic is given by the equation

$$M = \frac{\prod_{j=1}^{m}|C_j|^{(n_j-1)/2}}{|C|^{(n-m)/2}} \tag{4.20}$$

where

n_j is the size of the jth sample,

$|C_j|$ is the determinant of the sample covariance matrix for the jth sample as defined in Section 2.7,

$|C|$ is the determinant of the *pooled covariance matrix*

$$C = \frac{\sum_{j=1}^{m}(n_j-1)C_j}{n-m}, \text{ and}$$

$n = \sum n_j$ is the total number of observations.

Large values of M provide evidence that the samples are not from populations with the same covariance matrix. An approximate F-test for whether an observed M-value is significantly large is provided by calculating

$$F = -2 \cdot b \log_e(M) \tag{4.21}$$

and finding the probability of a value this large or larger for an F distribution with v_1 and v_2 df, where

$$v_1 = \tfrac{1}{2}p(p+1)(m-1), \ v_2 = \frac{v_1+2}{c_2-c_1^2}, \ b = \frac{1-c_1-\frac{v_1}{v_2}}{v_1},$$

$$c_1 = \frac{(2p^2+3p-1)\left(\sum_{j=1}^{m}\frac{1}{n_j-1}-\frac{1}{n-m}\right)}{6(p+1)(m-1)}, \text{ and}$$

$$c_2 = \frac{(p-1)(p+2)\left(\sum_{j=1}^{m}\frac{1}{(n_j-1)^2}-\frac{1}{(n-m)^2}\right)}{6(m-1)}.$$

The F approximation of Equation 4.21 is only valid for $c_2 > c_1$. If $c_2 \leq c_1$, then an alternative approximation is used. In this alternative case, the F-value is calculated to be

$$F = -\frac{2b_1 \cdot \nu_2 \cdot \log_e(M)}{\nu_1 + 2b_1 \cdot \log_e(M)} \qquad (4.22)$$

where $b_1 = \dfrac{1 - c_1 - \dfrac{2}{\nu_2}}{\nu_2}$. This is tested against the F distribution with ν_1 and ν_2 df to see whether it is significantly large.

Box's test is sensitive to deviations from normality in the distribution of the variables. For this reason, robust alternatives to Box's test are recommended here, these being generalizations of what was suggested for the two-sample situation. Thus, absolute deviations from sample medians can be calculated for the data in m samples. For a single variable, these can be treated as the observations for a one-factor analysis of variance. A significant F-ratio is then evidence that the samples come from populations with different mean deviations, that is, populations with different covariance matrices. With more than one variable, any of the four tests described in the last section can be applied to the transformed data, and a significant result indicates that the covariance matrix is not constant for the m populations sampled.

Alternatively, the variables can be standardized to have unit variances for all the data lumped together and d-values calculated using Equation 4.13. These d-values can then be analyzed by a one-factor analysis of variance. This generalizes Van Valen's test, which was suggested for comparing the variation in two multivariate samples. A significant F-ratio from the analysis of variance indicates that some of the m populations sampled are more variable than others. As in the two-sample situation, this test is only really appropriate when some samples may be more variable than others for all the measurements being considered.

4.8.1 Example 4.4: Comparison of Samples of Egyptian Skulls

The data shown in Table 1.2 has four measurements on male Egyptian skulls for five samples from various past ages.

A one-factor analysis of variance on the first variable, maximum breadth, provides $F = 5.95$, with 4 and 145 df (Table 4.4). Here, $p < 0.001$, and hence, there is clear evidence that the population mean changed with time. For the other three variables, there is evidence that the population means changed with time for two of the other three variables: basibregmatic height, $F = 2.45$ (significant at the 5% level); basialveolar length, $F = 8.31$ (significant at the 0.1% level); and nasal height, $F = 1.51$ (not significant). Hence, there is evidence that the population mean changed with time for the first three variables.

Next, consider the four variables together. If the five samples are combined, then the matrix of sums of squares and products for the 150 observations, calculated using Equation 4.15, is

$$\mathbf{T} = \begin{bmatrix} 3563.89 & -222.81 & -615.16 & 426.73 \\ -222.81 & 3635.17 & 1046.28 & 346.47 \\ -615.16 & 1046.28 & 4309.26 & -16.40 \\ 426.73 & 346.47 & -16.40 & 1533.33 \end{bmatrix}$$

for which the determinant is $|\mathbf{T}| = 7.306 \times 10^{13}$. Also, the within-sample matrix of sums of squares and cross products is found from Equation 4.16 to be

$$\mathbf{W} = \begin{bmatrix} 3061.07 & 5.33 & 11.47 & 291.30 \\ 5.33 & 3405.27 & 754.00 & 412.53 \\ 11.47 & 754.00 & 3505.97 & 164.33 \\ 291.30 & 412.53 & 164.33 & 1472.13 \end{bmatrix}$$

for which the determinant is $|\mathbf{W}| = 4.848 \times 10^{13}$. Wilks' lambda statistic (Equation 4.14) is therefore

$$\Lambda = |\mathbf{W}|/|\mathbf{T}| = 0.6636.$$

With $p = 4$ variables, $m = 5$ samples, and $n = 150$ observations in total, we have (using the notation in Table 4.5) that

$$df_1 = p(m-1) = 16$$

$$w = n - 1 - (p+m)/2 = 150 - 1 - (4+5)/2 = 144.5$$

$$t = \left\{ (df_1^2 - 4)/[p^2 + (m-1)^2 - 5] \right\}^{1/2} = \left\{ (16^2 - 4)/[4^2 + (5-1)^2 - 5] \right\}^{1/2} = 3.055$$

and

$$df_2 = w \cdot t - df_1/2 + 1 = 144.5 \times 3.055 - 16/2 + 1 = 434.5$$

The F-statistic is then

$$F = \left[(1 - \Lambda^{1/t})/\Lambda^{1/t} \right](df_2/df_1)$$
$$= \left[(1 - 0.6636^{1/3.055})/0.6636^{1/3.055} \right](434.5/16) = 3.90$$

with 16 and 434.5 df. As $p < 0.001$, there is clear evidence that the vector of mean values of the four variables changed with time.

The maximum root of the matrix $\mathbf{W}^{-1}\mathbf{B}$ is $\lambda_1 = 0.4251$ for Roy's maximum root test. From Table 4.5:

$$df_1 = d = \max(p, m-1) = \max(4, 5-1) = 4;$$
$$df_2 = n - m - d - 1 = 150 - 5 - 4 - 1 = 140.$$

The corresponding approximate F-statistic is

$$F = (df_2 / df_1)\lambda_1 = (140 / 4)0.4251 = 14.88$$

with 4 and 140 df, using the equations given in the table for the df. This, again, is very significantly large (p < 0.001).

Pillai's trace statistic (Equation 4.18) is $V = 0.3533$. The approximate F-statistic in this case is

$$F = (n - m - p + s)V / [d(s - V)] = 3.51$$

with $s \cdot d = 16$ and $s(n - m - p + s) = 580$ df, using the equations given in Table 4.5. This is another very significant result (p < 0.001).

Finally, for the tests on the mean vectors, the Lawley–Hotelling trace statistic (Equation 4.19) has the value $U = 0.4818$. It is found using the equations in Table 4.5 that the intermediate quantities that are needed are $s = 4$, $A = -0.5$, and $B = 70$ so that the df for the F-statistic are $df_1 = s(2A + s + 1) = 16$ and $df_2 = 2(sB + 1) = 562$. The F-statistic is then

$$F = df_2 \cdot U / (s \cdot df_1) = (562 \times 0.4818) / (4 \times 16) = 4.23.$$

Yet again, this is a very significant result (p < 0.001).

To compare the variation in the five samples, first consider Box's test. Equation 4.20 gives $M = 2.869 \times 10^{-11}$. The equations in the last section then give $b = 0.0235$,

$$F = -2 \cdot b \log_e(M) = 1.14,$$

with $\nu_1 = 40$ and $\nu_2 = 46,379$ df. This is not at all significantly large (p = 0.250), so this test gives no evidence that the covariance matrix changed with time.

Box's test is reasonable with this set of data because body measurements tend to have distributions that are close to normal. However, robust tests can also be carried out. It is a straightforward matter to transform the data into absolute deviations from sample medians for Levene-type tests. Analysis of variance then shows no significant difference between the sample means of the transformed data for any of the four variables considered individually. Also, none of the multivariate tests summarized in Table 4.5 give a result that is anything like significant at the 5% level for all the variables taken together.

Although there is very strong evidence that mean values changed with time for the four variables being considered, there is no evidence at all that the variation changed.

4.9 Computer Programs

The tests for multivariate normal data from this chapter are available in standard statistical computer packages, although many packages will be missing one or two of them. The tests can also be carried out using the R code described in the supplementary material for this chapter. The results of tests based on F distribution approximations may vary to some extent from one program to the next because of the use of different approximations. On the other hand, for the robust tests on variances, some or all of the calculations may need to be done in a spreadsheet or in R.

In this chapter, we restricted attention to two or more multivariate samples being compared to see whether they seem to come from populations with different mean vectors (one-factor MANOVA) or with different covariance matrices. More complicated examples involve samples being classified on the basis of several factors, giving generalizations of ordinary analysis of variance (ANOVA) and one-factor MANOVA. Most statistical packages allow the general MANOVA calculations to be performed.

Exercises

1.

    a. **R.** Verify all statistics and t-tests displayed in Table 4.1, by using any of the two methods described in Example 4.1 for the `t.test` function.

    b. **R.** Make use of the code and results presented in the addendum to Example 4.1 included in the supplementary material for Chapter 4. Expand Table 4.1 with two additional columns: the first containing the Bonferroni-corrected p-values and the second with the False Discovery Rate correction (Benjamini and Hochberg, 1995). Explain the differences between these methods for testing differences of univariate means.

    c) **R.** Use the source function `Levenetests2s.mv` to produce Bonferroni-corrected p-values for the comparison of variation between survivor and nonsurvivor sparrows, via univariate one-sided Levene tests, as described in Example 4.2. Is the variation of length of the humerus still significantly lower for survivors than for nonsurvivors?

2. Example 1.4 concerned the comparison between prehistoric dogs from Thailand and six other related animal groups in terms of mean mandible measurements. Table 4.6 shows some further data for the comparison of these groups, which are part of the more extensive data discussed in the paper by Higham et al. (1980).

a. Test for significant differences between the five species in terms of the mean values and the variation in the nine variables. Test both for overall differences and for differences between the prehistoric Thai dogs and each of the other groups singly. What conclusion do you draw with regard to the similarity between prehistoric Thai dogs and the other groups?

b. Is there evidence of differences between the size of males and females of the same species for the first four groups?

c. Using a suitable graphical method, compare the distribution of the nine variables for the prehistoric and modem Thai dogs.

TABLE 4.6

Values for Nine Mandible Measurements for Samples of Five Canine Groups, with Sex Information Included

	X_1	X_2	X_3	X_4	X_5	X_6	X_7	X_8	X_9	Sex
Modern dogs from Thailand										
1	123	10.1	23	23	19	7.8	32	33	5.6	Male
2	127	9.6	19	22	19	7.8	32	40	5.8	Male
3	121	10.2	18	21	21	7.9	35	38	6.2	Male
4	130	10.7	24	22	20	7.9	32	37	5.9	Male
5	149	12.0	25	25	21	8.4	35	43	6.6	Male
6	125	9.5	23	20	20	7.8	33	37	6.3	Male
7	126	9.1	20	22	19	7.5	32	35	5.5	Male
8	125	9.7	19	19	19	7.5	32	37	6.2	Male
9	121	9.6	22	20	18	7.6	31	35	5.3	Female
10	122	8.9	20	20	19	7.6	31	35	5.7	Female
11	115	9.3	19	19	20	7.8	33	34	6.5	Female
12	112	9.1	19	20	19	6.6	30	33	5.1	Female
13	124	9.3	21	21	18	7.1	30	36	5.5	Female
14	128	9.6	22	21	19	7.5	32	38	5.8	Female

(Continued)

TABLE 4.6   (Continued)

	X_1	X_2	X_3	X_4	X_5	X_6	X_7	X_8	X_9	Sex
15	130	8.4	23	20	19	7.3	31	40	5.8	Female
16	127	10.5	25	23	20	8.7	32	35	6.1	Female
Golden jackals										
1	120	8.2	18	17	18	7.0	32	35	5.2	Male
2	107	7.9	17	17	20	7.0	32	34	5.3	Male
3	110	8.1	18	16	19	7.1	31	32	4.7	Male
4	116	8.5	20	18	18	7.1	32	33	4.7	Male
5	114	8.2	19	18	19	7.9	32	33	5.1	Male
6	111	8.5	19	16	18	7.1	30	33	5.0	Male
7	113	8.5	17	18	19	7.1	30	34	4.6	Male
8	117	8.7	20	17	18	7.0	30	34	5.2	Male
9	114	9.4	21	19	19	7.5	31	35	5.3	Male
10	112	8.2	19	17	19	6.8	30	34	5.1	Male
11	110	8.5	18	17	19	7.0	31	33	4.9	Female
12	111	7.7	20	18	18	6.7	30	32	4.5	Female
13	107	7.2	17	16	17	6.0	28	35	4.7	Female
14	108	8.2	18	16	17	6.5	29	33	4.8	Female
15	110	7.3	19	15	17	6.1	30	33	4.5	Female
16	105	8.3	19	17	17	6.5	29	32	4.5	Female
17	107	8.4	18	17	18	6.2	29	31	4.3	Female
18	106	7.8	19	18	18	6.2	31	32	4.4	Female
19	111	8.4	17	16	18	7.0	30	34	4.7	Female
20	111	7.6	19	17	18	6.5	30	35	4.6	Female
Cuons										
1	123	9.7	22	21	20	7.8	27	36	6.1	Male
2	135	11.8	25	21	23	8.9	31	38	7.1	Male
3	138	11.4	25	25	22	9.0	30	38	7.3	Male
4	141	10.8	26	25	21	8.1	29	39	6.6	Male
5	135	11.2	25	25	21	8.5	29	39	6.7	Male
6	136	11.0	22	24	22	8.1	31	39	6.8	Male
7	131	10.4	23	23	23	8.7	30	36	6.8	Male
8	137	10.6	25	24	21	8.3	28	38	6.5	Male
9	135	10.5	25	25	21	8.4	29	39	6.9	Male
10	131	10.9	25	24	21	8.5	29	35	6.2	Female
11	130	11.3	22	23	21	8.7	29	37	7.0	Female

(Continued)

TABLE 4.6   (Continued)

	X_1	X_2	X_3	X_4	X_5	X_6	X_7	X_8	X_9	Sex
12	144	10.8	24	26	22	8.9	30	42	7.1	Female
13	139	10.9	26	23	22	8.7	30	39	6.9	Female
14	123	9.8	23	22	20	8.1	26	34	5.6	Female
15	137	11.3	27	26	23	8.7	30	39	6.5	Female
16	128	10.0	22	23	22	8.7	29	37	6.6	Female
17	122	9.9	22	22	20	8.2	26	36	5.7	Female
Indian wolves										
1	167	11.5	29	28	25	9.5	41	45	7.2	Male
2	164	12.3	27	26	25	10.0	42	47	7.9	Male
3	150	11.5	21	24	25	9.3	41	46	8.5	Male
4	145	11.3	28	24	24	9.2	36	41	7.2	Male
5	177	12.4	31	27	27	10.5	43	50	7.9	Male
6	166	13.4	32	27	26	9.5	40	47	7.3	Male
7	164	12.1	27	24	25	9.9	42	45	8.3	Male
8	165	12.6	30	26	25	7.7	40	43	7.9	Male
9	131	11.8	20	24	23	8.8	38	40	6.5	Female
10	163	10.8	27	24	24	9.2	39	48	7.0	Female
11	164	10.7	24	23	26	9.5	43	47	7.6	Female
12	141	10.4	20	23	23	8.9	38	43	6.0	Female
13	148	10.6	26	21	24	8.9	39	40	7.0	Female
14	158	10.7	25	25	24	9.8	41	45	7.4	Female
Prehistoric Thai dogs										
1	112	10.1	17	18	19	7.7	31	33	5.8	Unknown
2	115	10.0	18	23	20	7.8	33	36	6.0	Unknown
3	136	11.9	22	25	21	8.5	36	39	7.0	Unknown
4	111	9.9	19	20	18	7.3	29	34	5.3	Unknown
5	130	11.2	23	27	20	9.1	35	35	6.6	Unknown
6	125	10.7	19	26	20	8.4	33	37	6.3	Unknown
7	132	9.6	19	20	19	9.7	35	38	6.6	Unknown
8	121	10.7	21	23	19	7.9	32	35	6.0	Unknown
9	122	9.8	22	23	18	7.9	32	35	6.1	Unknown
10	124	9.5	20	24	19	7.6	32	37	6.0	Unknown

Note: The variables are X_1 = length of mandible, X_2 = breadth of mandible below 1st molar, X_3 = breadth of articular condyle, X_4 = height of mandible below first molar, X_5 = length of first molar, X_6 = breadth of first molar, X_7 = length of first to third molar inclusive (first to second for cuon), X_8 = length from first to fourth premolar inclusive, and X_9 = breadth of lower canine, all measured in millimeters.

References

Benjamini, Y., and Hochberg, Y. (1995). Controlling the false discovery rate: a practical and powerful approach to multiple testing. *Journal of the Royal Statistical Society Series B* 57: 289–300.

Higham, C.F.W., Kijngam, A., and Manly, B.F.J. (1980). An analysis of prehistoric canid remains from Thailand. *Journal of Archaeological Science* 7: 149–165.

Kres, H. (1983). *Statistical Tables for Multivariate Analysis*. New York: Springer.

Levene, H. (1960). Robust tests for equality of variance. In Olkin, I., Ghurye, S.G., Hoeffding, W., Madow, W.G., and Mann, H.B. (eds). *Contributions to Probability and Statistics*, pp. 278–292. Palo Alto, CA: Stanford University Press.

Manly, B.F.J. (2009). *Statistics for Environmental Science and Management*. 2nd Edn. Boca Raton, FL: Chapman and Hall/CRC.

Manly, B.F.J., and Francis, R.I.C.C. (2002). Testing for mean and variance differences with samples from distributions that may be non-normal with unequal variances. *Journal of Statistical Computation and Simulation* 72: 633–646.

Schultz, B. (1983). On Levene's test and other statistics of variation. *Evolutionary Theory* 6: 197–203.

Seber, G.A.F. (2004). *Multivariate Observations*. New York: Wiley.

Van Valen, L. (1978). The statistics of variation. *Evolutionary Theory* 4: 33–43. (Erratum *Evolutionary Theory* 4: 202.)

Welch, B.L. (1951). On the comparison of several mean values: an alternative approach. *Biometrika* 38: 330–336.

Yao, Y. (1965). An approximate degrees of freedom solution to the multivariate Behrens-Fisher problem. *Biometrika* 52: 139–147.

Appendix: Tests of Significance in R

A.1 Univariate Two-Sample t-test in R

The two-sample *t*-test can be conducted with the `t.test` function. By default, a two-sided test and that the within-sample variances are not equal are assumed. Therefore, under the assumption of common variances, it is necessary to write

```
t.test(vector1, vector2, var.equal = TRUE)
```

where `vector1` and `vector2` are the numeric vectors for Samples 1 and 2 of a single numeric response variable, respectively. An alternative method to perform a `t.test` makes use of a formula:

```
t.test(y ~ x, data = df, ...)
```

`y` is a single numeric response variable, and `x` is a factor with two levels giving the corresponding samples or groups; both variables are contained in the data frame `df`.

Alternatively, as an aid to users of this book, function `ttests2s.mv` can be accessed from source code **ttests2s.mv.R** (downloadable from the book's electronic resources site) to run several two-sample t-tests with significance levels corrected by multiple testing with any of the adjustment methods for multiple comparisons implemented in the function `p.adjust`. We encourage to read the help file for `p.adjust` in order to know the adjustment methods available. The document containing supplementary materials for Chapter 4 provide an example of the execution of `ttests2s.mv`, applied to Example 4.1. See also Exercise 1, in this chapter.

A.2 Multivariate Two-Sample Tests with Hotelling's T^2

The results of Hotelling's T^2 test are easily computed via the `hotelling.test` function, contained in the package Hotelling (Curran and Hersh, 2021).

```
library(Hotelling)
```

```
hotelling.test(formula)
```

Here, `formula` is an expression in which the response variables, separated by sum (+) signs, are followed by a tilde (~) and a factor vector with two levels. As an example, if Y1, Y2, and Y3 are three response variables in a data frame, and X is a two-level grouping variable, then `hotelling.test(Y1 + Y2 + Y3 ~ X)` invokes Hotelling's test for the comparison of the two multivariate means.

An alternative R-function for the T^2 test is provided in the source file **Hotelling.mat.R**. The syntax of the function is

```
Hotelling.mat(x, group, level1)
```

where x is a data frame with $p + 1$ columns, being p of them numeric response variables, and the remaining column, `group`, is a two-factor variable, written without quotes. Finally, `level1` is a character string specifying the first level of interest in `group`. `Hotelling.mat` extends the output generated by `hotelling.test`, by offering a summary method (option `long = TRUE`) that optionally prints mean vectors and covariance matrices used in the intermediate calculations of the Hotelling's T^2 statistic.

A.3. Comparison of Variation of Two Samples for the Univariate Case

R offers the two-sample variance tests described in Section 4.5, the F-ratio test and Levene's test, where the latter requires the package `car` (Fox and Weisberg, 2019) to be loaded. The F-test is invoked via the `var.test` function, its syntax being similar to the `t.test()` function:

```
var.test(vector1, vector2)
```

or

```
var.test(y ~ x, data = df, ...)
```

being `y` a numeric response variable and `x` a factor with two levels giving the corresponding samples or groups. By default, the alternative hypothesis for `var.test` is two-sided.

On the other hand, Levene's test, as implemented in function `leveneTest` from package `car`, does not restrict the grouping variable to two levels only. The code for this function is

```
library(car)

leveneTest(response ~ x, data = df)
```

The first argument in the `leveneTest` function is a vector containing the `response` variable, and `x` is the factor defining groups (with at least two levels). As `leveneTest` is not restricted to two levels only, it produces an F statistic, instead of a t statistic. In addition, the number of degrees of freedom for a two-level factor is equal to 1 thus, the relation $F = t^2$ holds. Therefore, the F and t-tests are equivalent. However, `leveneTest` is a two-sided test; thus, the significance level produced by `leveneTest` for one-sided Levene's tests must be halved. The R-source file **Levenetests2s.mv.R** offers an alternative function `Levenetests2s.mv` that implements the methodology described in this book. This function is based on individual t-statistics computed to test the variation of absolute deviations from the medians for more than one variable and allows for the computation of p-values for one or two-sided tests directly, and it is possible to choose a method of p-value adjustment for multiple testing.

A.4 Comparison of Variation of Two Multivariate Samples

The R commands required for testing the multivariate variation of two groups as outlined in Section 4.6 make use of basic vector functions like Levene's test based on Hotelling's T^2. The source file **LeveneT2.R**, downloadable from the website of electronic resources for this book, can be executed to compare two mean vectors of absolute deviations from medians (*ADMs*) using Hotelling's T^2 test, under a Levene's test strategy. Another downloadable source file, **VanValen.R**, implements the function `VanValen` to produce Van Valen's test. Data standardization is computed with the base function `scale`. Scripts executing functions `LeveneT2` and `VanValen`, applied to the Bumpus' sparrows data for Example 4.2, are described in the supplementary material document for Chapter 4.

A.5 Comparison of Several Multivariate Means: One-Factor MANOVA

A preamble to one-factor MANOVA, the (univariate) one-factor ANOVA, is briefly described at the beginning of Section 4.7. The calculations for univariate analysis of variance (summarized in Table 4.4) can be performed in R with `lm` (linear models) or `aov` (analysis of variance). The approach followed in this book indicates choosing `aov`, as it is closely related to the traditional language of the analysis of variance rather than that of linear models. The syntax for the one-factor ANOVA considered in Section 4.7 is

```
aov.object <- aov(y ~ x, data = df)

summary(aov.object)
```

where `y` is a single response variable and `x` is a factor, and both variables contained in data frame `df`.

Calculations for one-factor MANOVA can be easily produced in R with the function `manova`, which is an extension of the corresponding univariate function `aov`. In fact, `manova` calls `aov` for computations, but supplementary information is also computed and made available as output, including sums of squares and cross products matrices, eigenvalues, and all test statistics listed in Table 4.5. This is described by saying that `manova` produces a different object, one of class `manova` (see more details in the R documentation). The simplest call to a one-way MANOVA is

```
manova.obj <- manova(mat ~ x, data = df)
```

Here, the analyzed response variables are contained in matrix `mat`, taken from the data frame `df`, and the samples to be compared are levels of factor `x`. Once the output of the `manova` function is assigned to the object `manova.obj`, any of the test statistics available as summaries of the MANOVA can be invoked with `summary`. Options for the `test =` argument are `"Wilks"`, `"Pillai"` (the default), `"Hotelling-Lawley"`, and `"Roy"`. The results of a Wilks' test can be obtained with

```
summary(manova.obj, test = "Wilks")
```

An additional source file **MANOVA.mat.R** is also included in the electronic resources website. This file includes a call to the function `MANOVA.mat` as an extension to the default `summary` method in `manova` (again with option `long = TRUE`), that allows to visualize the matrices **T**, **W** and **B** defined in Section 4.7.

A.6 Testing the Equality of Several Covariance Matrices

The simplest way to run Box's M-test is by means of the function `BoxM`, found in package `biotools` (da Silva et al., 2017; da Silva, 2021). This function uses a chi-squared approximation instead of the procedure described in Section 4.8, which is based on the F-statistic. Morrison (2005) gives the computational details for this chi-squared approximation, which is applicable whenever the number of groups and the number of variables do not exceed four or five, and the sample sizes per group are equal to or greater than 20. The F approximation is more suitable for greater numbers of groups and variables and smaller sample sizes per group. The formulae for both approximations are linked, so that it is not difficult to obtain the approximated F-statistic from the output generated by the function `BoxM`. An alternative function for Box's M-test,

`BoxM.F` is provided by the source file **BoxM.F.R** specifically written for the procedure based on an approximate *F*-test, described in Section 4.8. The `summary` method for `BoxM.F` offers extended (`long = TRUE`) output of matrices involved in the test. An example on the use of `BoxM` and `BoxM.F`, used in Example 4.3, is available in the supplementary material linked to the book.

References

Curran, J., and Hersh, T. (2021). *Hotelling: Hotelling's T^2 Test and Variants*. R package version 1.0-8. https://CRAN.R-project.org/package=Hotelling.

da Silva, A.R. (2021). *biotools: Tools for Biometry and Applied Statistics in Agricultural Science*. R package version 4.2. https://cran.r-project.org/package=biotools.

da Silva, A.R., Malafaia, G., and Menezes, I.P.P. (2017). biotools: an R function to predict spatial gene diversity via an individual-based approach. *Genetics and Molecular Research* 16. https://doi.org/10.4238/gmr16029655.

Fox, J., and Weisberg, S. (2019). *An R Companion to Applied Regression*. 3rd Edn. Thousand Oaks, CA: SAGE. https://socialsciences.mcmaster.ca/jfox/Books/Companion/.

Morrison, D. (2005). *Multivariate Statistical Methods*. 4th Edn. Pacific Grove, CA: Duxbury.

5

Measuring and Testing Multivariate Distances

5.1 Multivariate Distances

Many multivariate problems can be viewed in terms of distances between single observations, between samples of observations, or between populations of observations. For example, considering the data in Table 1.4 on mandible measurements of dogs, wolves, jackals, cuons, and dingos, it is reasonable to ask how far one of these groups is from the other six groups. The idea is that if two animals have similar mean mandible measurements, then they are close, whereas if they have rather different mean measurements, then they are distant from each other. Throughout this chapter, this is the concept of distance we use.

More than a few distance measures have been proposed and used in multivariate analyses. Here, only some of the most common ones will be mentioned. It is fair to say that measuring distances is a topic in which a certain amount of arbitrariness seems unavoidable.

One situation is that there are n objects being considered, with a number of measurements being taken on each and distances between the objects are computed and summarized in a distance matrix. A simple one to imagine is a matrix of geographic distances between pairs of a set of n locations on a map. The diagonal elements of this will be zero. An analogous concept applies in genetic analyses, wherein one can calculate the "genetic" distance between pairs of a collection of (say) species of plants or animals. For example, in Table 1.3, results are given for four environmental variables and six gene frequencies for 16 colonies of a butterfly. Two sets of distances, environmental and genetic, can, therefore, be calculated between the colonies. An interesting question then is whether there is a significant relationship between these two sets of distances. Mantel's test (Section 5.6) is useful in this context.

DOI: 10.1201/9781003453482-5

5.2 Distances between Individual Observations

To begin, consider the simplest case where there are n objects, each of which has values for p variables, $X_1, X_2, \ldots X_p$. The values for object i can then be denoted by $x_{i1}, x_{i2}, \ldots x_{ip}$. The problem is to measure the distance between these two objects. If there are only $p = 2$ variables, then the values can be plotted as shown in Figure 5.1. Pythagoras' theorem then says that the length d_{ij} of the line joining the point for object i to the point for object j (the Euclidean distance) is

$$d_{ij} = \left[\left(x_{i1} - x_{j1} \right)^2 + \left(x_{i2} - x_{j2} \right)^2 \right]^{1/2}.$$

With $p = 3$ variables, the values can be taken as the coordinates in space for plotting the positions of individuals i and j (Figure 5.2). Pythagoras' theorem then gives the distance between the two points to be

$$d_{ij} = \left[\left(x_{i1} - x_{j1} \right)^2 + \left(x_{i2} - x_{j2} \right)^2 + \left(x_{i3} - x_{j3} \right)^2 \right]^{1/2}.$$

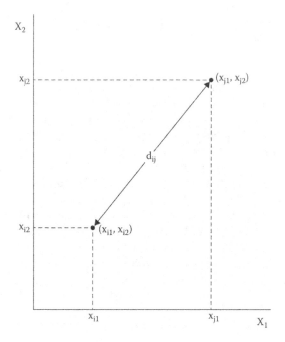

FIGURE 5.1
The Euclidean distance between objects i and j with $p = 2$ variables.

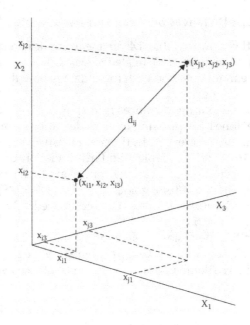

FIGURE 5.2
The Euclidean distance between objects i and j with $p = 3$ variables.

With more than three variables, it is not possible to use variable values as the coordinates for physically plotting points. However, the two- and three-variable cases suggest that the generalized Euclidean distance

$$d_{ij} = \left[\sum_{k=1}^{p} \left(x_{ik} - x_{jk} \right)^2 \right]^{1/2} \tag{5.1}$$

may serve as a satisfactory measure for many purposes with p variables. From the form of Equation 5.1, it is clear that if one of the variables measured is much more variable than the others, then this will dominate the calculation of distances. For example, to take an extreme case, suppose that n men are being compared and that X_1 is their stature and the other variables are tooth dimensions, with all the measurements being in millimeters. Stature differences will then be in the order of perhaps 50 or 100 mm, while tooth dimension differences will be in the order of 1 or 2 mm. The simple calculations of d_{ij} will then provide distances between individuals that are essentially stature differences only, with tooth differences having a negligible effect.

In practice, it is usually desirable for all variables to have about the same influence on the distance calculation. This can be achieved by a preliminary scaling that involves dividing each variable by its standard deviation for the n individuals being compared.

5.2.1 Example 5.1: Distances between Dogs and Related Species

Consider again the data in Table 1.4 for mean mandible measurements of seven groups of Thai dogs and related species. It may be recalled from Chapter 1 that the main question with these data is how the prehistoric Thai dogs relate to the other groups.

The first step in calculating distances is to standardize the measurements. Here, this will be done by expressing them as deviations from means in units of standard deviations. For example, the first measurement X_1 (breadth) has a mean of 10.49 mm and a sample standard deviation of 1.70 mm for the seven groups. The standardized variable values are then calculated as: modern dog, $(9.7 - 10.49)/1.7 = -0.46$; golden jackal, $(8.1 - 10.49)/1.7 = -1.41, \ldots$ prehistoric dog, $(10.3 - 10.49)/1.7 = -0.11$. Standardized values for all the variables are shown in Table 5.1.

Using Equation 5.1, the distances shown in Table 5.2 have been calculated from the standardized variables. Clearly, prehistoric dogs are rather similar to modern dogs in Thailand, because the distance of 0.66 between these two

TABLE 5.1

Standardized Variable Values Calculated from the Original Data in Table 1.4

Group	X_1	X_2	X_3	X_4	X_5	X_6
Modern dog	−0.46	−0.46	−0.68	−0.69	−0.45	−0.57
Golden jackal	−1.41	−1.79	−1.04	−1.29	−0.80	−1.21
Chinese wolf	1.78	1.48	1.70	1.80	1.55	1.50
Indian wolf	0.60	0.55	0.96	0.69	1.17	0.88
Cuon	0.13	0.31	−0.04	0.00	−1.10	−0.37
Dingo	−0.52	0.03	−0.13	−0.17	0.03	0.61
Prehistoric dog	−0.11	−0.12	−0.78	−0.34	−0.41	−0.83

TABLE 5.2

Euclidean Distances Between Seven Canine Groups

	Modern dog	Golden jackal	Chinese wolf	Indian wolf	Cuon	Dingo	Prehistoric dog
Modern dog	—						
Golden jackal	1.91	—					
Chinese wolf	5.38	7.12	—				
Indian wolf	3.38	5.06	2.14	—			
Cuon	1.51	3.19	4.57	2.91	—		
Dingo	1.56	3.18	4.21	2.20	1.67	—	
Prehistoric dog	0.66	2.39	5.12	3.24	1.26	1.71	—

groups is by far the smallest distance in the whole table. Higham et al. (1980) concluded from a more complicated analysis that modern and prehistoric dogs are indistinguishable.

5.3 Distances between Populations and Samples

Several measures have been proposed for the distance between multivariate populations when information is available on the means, variances, and covariances of the populations. Here, two measures will be considered.

Suppose that two or more populations are available, and the multivariate distributions in these populations are known for p variables $X_k; k = 1, ..., p$. Let the mean of X_k in population i be μ_{ki}, and assume that the variance of X_k is V_k in all the populations. Then, Penrose (1953) proposed the relatively simple measure

$$P_{ij} = \sum_{k=1}^{p} \frac{\left(\mu_{ki} - \mu_{kj}\right)^2}{pV_k} \tag{5.2}$$

for the distance between population i and population j.

A disadvantage of Penrose's measure is that it does not consider the correlations between the p variables. This means that when two variables are measuring essentially the same thing, and hence are highly correlated, they still individually both contribute about the same amount to population distances as a third variable that is uncorrelated with all other variables.

A measure that does take into account the correlations between variables is the Mahalanobis (1948) distance

$$D_{ij}^2 = \sum_{r=1}^{p} \sum_{s=1}^{p} \left(\mu_{ri} - \mu_{rj}\right) v_{(rs)} \left(\mu_{si} - \mu_{sj}\right) \tag{5.3}$$

where $v_{(rs)}$ is the element in row r and column s of the inverse of the population covariance matrix for the p variables. This is a quadratic form that can also be written as

$$D_{ij}^2 = (\mu_i - \mu_j)' \mathbf{V}^{-1} (\mu_i - \mu_j)$$

where:

μ_i is the population mean vector for population i and

$\mathbf{V}$ is the population covariance matrix.

This measure requires that $\mathbf{V}$ is the same for all populations.

A Mahalanobis distance is also often used to measure the distance of a single multivariate observation from the center of its population. If $x_1, x_2, ..., x_p$ are the values of $X_1, X_2, ..., X_p$ for the individual, with corresponding population means of $\mu_1, \mu_2, ..., \mu_p$, then this distance is

$$D^2 = \sum_{r=1}^{p} \sum_{s=1}^{p} (x_r - \mu_r) v_{(rs)} (x_s - \mu_s) = (\mathbf{x} - \mathbf{\mu})' \mathbf{V}^{-1} (\mathbf{x} - \mathbf{\mu}) \qquad (5.4)$$

where

$\mathbf{x} = (x_1, x_2, ..., x_p)$,

μ is the population mean vector,

$\mathbf{V}$ is the population covariance matrix, and

$v_{(rs)}$ is the element in the row r and column s of the inverse of V.

The value of D^2 can be thought of as a multivariate residual for the observation $\mathbf{x}$, that is, a measure of how far $\mathbf{x}$ is from the center of the distributions of all values, taking into account all the variables being considered and their covariances. An important result is that the values of will follow a chi-squared distribution with p degrees of freedom (df) if $\mathbf{x}$ comes from a multivariate normal distribution. A significantly large value of D^2 means that the corresponding observation is either (a) a genuine but unlikely record, (b) an observation from another distribution, or (c) a record containing some mistake. Observations with large Mahalanobis residuals should, therefore, be examined to see whether they have just been recorded wrongly.

Equations 5.2 through 5.4 can be used with sample data if estimates of population means, variances, and covariances are used in place of true values. In that case, the covariance matrix $\mathbf{V}$ involved in Equations 5.3 and 5.4 should be replaced with the pooled estimate from all the samples available, as defined in Section 4.8 for Box's M-test.

In principle the Mahalanobis distance is superior to the Penrose distance, because it uses information on covariances. However, this advantage is only present when the covariances are accurately known. When covariances can only be estimated rather poorly from small samples, it is probably best to use the simpler Penrose measure. It is difficult to say precisely what a small sample means in this context. Certainly, there should be no problem with using Mahalanobis distances based on a covariance matrix estimated with a total sample size of 100 or more.

5.3.1 Example 5.2: Distances between Samples of Egyptian Skulls

For the five samples of male Egyptian skulls shown in Table 1.2, the mean vectors and covariance matrices are shown in Table 5.3, as is the pooled covariance matrix. Although the five sample covariance matrices appear to differ somewhat, it has been shown in Example 4.3 that the differences are not significant.

Penrose's distance measures of Equation 5.2 can now be calculated between each pair of samples. There are $p = 4$ variables with variances that are estimated by $V_1 = 21.111$, $V_2 = 23.485$, $V_3 = 24.179$, and $V_4 = 10.153$, these being the diagonal terms in the pooled covariance matrix (Table 5.3). The sample mean values given in the vectors $\bar{x}_1$ to $\bar{x}_5$ are estimates of the population means μ_1 to μ_5, respectively. For example, the distance between Sample 1 and Sample 2 is calculated as

$$P_{12} = \frac{(131.37-132.37)^2}{4\times21.11} + \frac{(133.60-132.70)^2}{4\times23.49} + \frac{(99.17-99.07)^2}{4\times24.18} + \frac{(50.53-90.23)^2}{4\times10.15}$$
$$= 0.023.$$

This only has meaning in comparison with the distances between the other pairs of samples, as shown in Table 5.4a.

Recall from Example 4.3 that the mean values change significantly from sample to sample. The Penrose distances show that the changes are cumulative over time, with the samples that are closest in time being relatively similar, whereas the samples that are far apart in time are more different.

Mahalanobis distances can be calculated from Equation 5.3, with the population covariance matrix V estimated by the pooled sample covariance matrix C. The matrix C is provided in Table 5.3, and the inverse is

$$C^{-1} = \begin{bmatrix} 0.0483 & 0.0011 & 0.0001 & -0.0099 \\ 0.0011 & 0.0461 & -0.0094 & -0.0121 \\ 0.0001 & -0.0094 & 0.0435 & -0.0022 \\ -0.0099 & -0.0121 & -0.0022 & 0.1041 \end{bmatrix}.$$

TABLE 5.3

The Samples' Mean Vectors and Covariance Matrices and the Pooled Sample Covariance Matrix for the Egyptian Skull Data

Sample (j)		Mean vector, $\bar{x}_j$	Sample covariance matrices			
			X_1	X_2	X_3	X_4
Early predynastic (1)	X_1	131.37	26.31	4.15	0.45	7.25
	X_2	133.60	4.15	19.97	-0.79	0.39
	X_3	99.17	0.45	-0.79	34.63	-1.92
	X_4	50.53	7.25	0.39	-1.92	7.64

(Continued)

TABLE 5.3 (Continued)

Sample (*j*)		Mean vector, $\bar{x}_j$	Sample covariance matrices			
			X_1	X_2	X_3	X_4
Late predynastic (2)	X_1	132.37	23.14	1.01	4.77	1.84
	X_2	132.70	1.01	21.60	3.37	5.62
	X_3	99.07	4.77	3.37	18.89	0.19
	X_4	50.23	1.84	5.62	0.19	8.74
12th–13th Dynasty (3)	X_1	134.47	26.31	4.15	0.45	7.25
	X_2	133.80	4.15	19.97	−0.79	0.39
	X_3	96.03	0.45	−0.79	34.63	−1.92
	X_4	50.57	7.25	0.39	−1.92	7.64
Ptolemaic period (4)	X_1	135.50	15.36	−5.53	−2.17	2.05
	X_2	132.30	−5.53	26.36	8.11	6.15
	X_3	94.53	−2.17	8.11	21.09	5.33
	X_4	51.97	2.05	6.15	5.33	7.96
Roman period (5)	X_1	136.17	28.63	−0.23	−1.88	−1.99
	X_2	130.33	−0.23	24.71	11.72	2.15
	X_3	93.50	−1.88	11.72	25.57	0.40
	X_4	51.37	−1.99	2.15	0.40	13.83
			Pooled Covariance Matrix			
			21.111	0.037	0.079	2.009
			0.037	23.485	5.200	2.845
			0.079	5.200	24.179	1.133
			2.009	2.845	1.133	10.153

Using this inverse and the sample means gives the Mahalanobis distance from Sample 1 to Sample 2 to be

$$
\begin{aligned}
D_{12}^2 = &\, (131.37 - 132.37) \cdot 0.0483 \cdot (131.37 - 132.37) + \\
&\, (131.37 - 132.37) \cdot 0.0011 \cdot (133.60 - 132.70) + \ldots + \\
&\, (50.53 - 50.23) \cdot (-0.0022) \cdot (99.17 - 99.07) + \\
&\, (50.53 - 50.23) \cdot 0.1041 \cdot (50.53 - 50.23) \\
= &\, 0.091.
\end{aligned}
$$

Calculating the other distances between samples in the same way provides the distance matrix shown in Table 5.4b.

A comparison between these and the Penrose distances shows a very good agreement. The Mahalanobis distances are three to four times as large as the Penrose distances. However, the relative distances between samples are

TABLE 5.4

Penrose and Mahalanobis Distances between Pairs of Samples of Egyptian Skulls

	Early predynastic	Late predynastic	12th-13th Dynasty	Ptolemaic period	Roman period
		(a) Penrose distances			
Early predynastic	—				
Late predynastic	0.023	—			
12th–13th Dynasty	0.216	0.163	—		
Ptolemaic period	0.493	0.404	0.108	—	
Roman period	0.736	0.583	0.244	0.066	—
		(b) Mahalanobis distances			
Early predynastic	—				
Late predynastic	0.091	—			
12th–13th Dynasty	0.903	0.729	—		
Ptolemaic period	1.881	1.594	0.443	—	
Roman period	2.697	2.176	0.911	0.219	—

almost the same for both measures. For example, the Penrose distance from the early predynastic sample to the Roman sample is $0.736/0.023 = 32.0$ times as great as the distance from the early to the late predynastic samples. The corresponding ratio for the Mahalanobis measure is $2.697/0.091 = 29.6$.

5.4 Multivariate Similarities

Instead of distances matrices, like those shown in Tables 5.2 and 5.3, several multivariate statistical procedures start with *similarity* matrices. *Similarity measures* can be interpreted as "inverted" distance measures: larger similarity values indicate higher similarity between two variables or two objects (less distance between them), and smaller values indicate less similarity (larger distance). Depending on the interest to analyze objects or variables, a *similarity matrix* **S** can be constructed for all pairs of objects k and l or for all pairs of variables i and j, thinking of the corresponding entries, S_{kl} or S_{ij}, as measurements of the amount of resemblance between them.

A correlation matrix (which measures the strength of a linear relationship) is an example of a similarity matrix between variables: larger values indicate higher correlation between the two corresponding variables, and smaller, less. The diagonal elements of this matrix will be unity. A distance matrix can be thought of as the complement to a similarity matrix: larger values imply "further apart," while smaller imply "closer." One could invert the values (take the reciprocal) in a correlation matrix **R** and call the result a distance matrix, **D**. Now larger values would imply the two relevant variables are "further apart" in the sense that their behaviors are more mutually independent, while smaller values correspond to two variables being "closer together" in their behavior.

In the last example, a similarity matrix **R** (the correlation matrix) was converted into a distance matrix by taking reciprocals; the inverse process is also possible. Given a distance matrix **D**, a similarity measure can be constructed by taking reciprocals to produce a similarity matrix, **S**. Each entry in **S** consist of a similarity measurement $s = 1/D$, computed from each entry D in **D**, that varies from zero to infinity. This calculation gives a similarity that ranges from infinity for two items that are no distance apart, to zero for two objects that are infinitely far apart. To avoid the manipulation of infinite similarities, one can alternatively define $s = 1/(1+D)$, a similarity measure ranging from 1 when $D = 0$, to 0 when D is infinite.

5.5 Presence-Absence Data

Suppose you are interested in the similarity between two plant species in terms of their distributions at ten sites. The data might then take the form shown in Table 5.5. Such data are often summarized, as shown in Table 5.6, as counts of the number of times that both species are present (a), only one species is present (b and c), or both species are absent (d). Thus, for the data in Table 5.5, $a = 3$, $b = 3$, $c = 3$, and $d = 1$.

In this situation, some of the commonly used similarity measures are

$$\text{the simple matching index} = \frac{a+d}{n},$$

$$\text{the Ochiai index} = \frac{a}{\sqrt{(a+b)(a+c)}},$$

$$\text{the Dice} - \text{Sorensen index} = \frac{2a}{2a+b+c}, \text{and}$$

TABLE 5.5

Presences and Absences of Two Species at Ten Sites (1 = presence, 0 = absence)

Site	1	2	3	4	5	6	7	8	9	10
Species 1	0	0	1	1	1	0	1	1	1	0
Species 2	1	1	1	1	0	0	0	0	1	1

TABLE 5.6

Presence and Absence Data Obtained for Two Species at n Sites

Species 1	Species 2		Total
	Present	Absent	
Present	a	b	$a+b$
Absent	c	d	$c+d$
Total	$a+c$	$b+d$	n

$$\text{the Jaccard index} = \frac{a}{a+b+c}$$

These all vary from zero (no similarity) to one (complete similarity) so that complementary distance measures can be calculated by subtracting the similarity indices from one. Symbolically, for any similarity s between cases varying from zero to one, it follows that $D = 1 - s$ is a measure of the distance between the cases being compared. This strategy in the definition of a distance measure using similarities is symmetrical: for any distance D between cases varying from zero to one, the quantity $s = 1 - D$ defines a measure of the similarity between those cases.

5.5.1 Example 5.3 Measuring Similarities between Two Plant Species

The similarity between Species 1 and 2 in Table 5.5, according to the four different indexes defined, is as follows: the simple matching index = 0.40, the Ochiai index = 0.50, the Dice-Sorensen index = 0.50, and the Jaccard index = 0.33. All measures suggest that similarities between these two species range from moderate to low.

It is debatable whether the number of joint absences (d) should be used in the calculation because of the danger of concluding that two species are similar simply because they are both absent from many sites. This is certainly a valid point in many situations and suggests that the simple matching index should be used with caution.

The four indices listed, and many other indices for binary data, are reviewed by Gower and Legendre (1986), while Jackson et al. (1989) compare the results of using different indices with various multivariate analyses of the presences and absences of 25 fish species in 52 lakes. Other studies comparing indices for binary data are those by Brusco et al. (2021) and Batyrshin (2016). Finally, the study by Cibulková et al. (2020) is recommended to readers interested in the comparison of those indices on the basis of their performance in cluster analysis, a topic covered in Chapter 9.

5.6 The Mantel Randomization Test

A useful test for comparing two distance or similarity matrices was introduced by Mantel (1967) as a solution to the problem of detecting space and time clustering of diseases, that is, whether cases of a disease that occur close in space also tend to be close in time.

To understand the nature of the procedure, the following simple example should be helpful. Suppose that four objects are being studied, and that two sets of variables have been measured for each of these. The first set of variables can then be used to construct a 4 × 4 matrix, where the entry in row i and column j is a measure of the distance between object i and object j based on these variables. The distance matrix might, for example, be

$$\mathbf{M} = \begin{bmatrix} m_{11} & m_{12} & m_{13} & m_{14} \\ m_{21} & m_{22} & m_{23} & m_{24} \\ m_{31} & m_{32} & m_{33} & m_{34} \\ m_{41} & m_{42} & m_{43} & m_{44} \end{bmatrix} = \begin{bmatrix} 0.0 & 1.0 & 1.4 & 0.9 \\ 1.0 & 0.0 & 1.1 & 1.6 \\ 1.4 & 1.1 & 0.0 & 0.7 \\ 0.9 & 1.6 & 0.7 & 0.0 \end{bmatrix}.$$

This is a symmetric matrix because, for instance, the distance from Object 2 to Object 3 must be the same as the distance from Object 3 to Object 2 (1.1 units). Diagonal elements are zero because these represent distances from objects to themselves.

The second set of variables can also be used to construct a matrix of distances between the objects. For the example, this will be taken as

$$\mathbf{E} = \begin{bmatrix} e_{11} & e_{12} & e_{13} & e_{14} \\ e_{21} & e_{22} & e_{23} & e_{24} \\ e_{31} & e_{32} & e_{33} & e_{34} \\ e_{41} & e_{42} & e_{43} & e_{44} \end{bmatrix} = \begin{bmatrix} 0.0 & 0.5 & 0.8 & 0.6 \\ 0.5 & 0.0 & 0.5 & 0.9 \\ 0.8 & 0.5 & 0.0 & 0.4 \\ 0.6 & 0.9 & 0.4 & 0.0 \end{bmatrix}.$$

Like $\mathbf{M}$, this is symmetric with zeros down the diagonal.

Mantel's test assesses whether the elements in $\mathbf{M}$ and $\mathbf{E}$ show some significant correlation. The test statistic that is used is sometimes the correlation between the corresponding elements of the two matrices (matching m_{11} with e_{11}, m_{12} with e_{12}, etc.), or the simpler sum of the products of these matched elements. For the general case of $n \times n$ matrices, the latter statistic is

$$Z = \sum_{i=2}^{n} \sum_{j=1}^{i-1} m_{ij} \cdot e_{ij} \tag{5.5}.$$

This statistic is calculated and compared with the distribution of Z that is obtained by taking the objects in a random order for one of the matrices, which is why it is called a randomization test.

For the randomization test, the matrix **M** can be left as it is. A random order can then be chosen for the objects for matrix **E**. For example, suppose that a random ordering of objects turns out to be 3,2,4,1. This then gives a randomized **E** matrix of

$$
E_R = \begin{bmatrix}
0.0 & 0.5 & 0.4 & 0.8 \\
0.5 & 0.0 & 0.9 & 0.5 \\
0.4 & 0.9 & 0.0 & 0.6 \\
0.8 & 0.5 & 0.6 & 0.0
\end{bmatrix}.
$$

The entry in row 1 and column 2 is 0.5, the distance between objects 3 and 2; the entry in row 1 and column 3 is 0.4, the distance between objects 3 and 4; and so on. A Z value can be calculated using **M** and E_R. Repeating this procedure using different random orders of the objects for E_R generates the randomized distribution of Z. A check can then be made to see whether the observed Z value is a typical value from this distribution.

The basic idea here is that if the two measures of distance are quite unrelated, then the matrix **E** will be just like one of the randomly ordered matrices E_R. Hence, the observed Z will be a typical randomized Z value. On the other hand, if the two distance measures have a positive correlation, then the observed Z will tend to be larger than values given by randomization. While unlikely, a negative correlation between distances will result in the observed Z value being low when compared with the randomized distribution.

With n objects, there are $n!$ different possible orderings of the object numbers. There are therefore $n!$ possible randomizations of the elements of **E**, some of which might give the same Z values. Hence, in our example with four objects, the randomized Z-distribution has $4! = 24$ equally likely values. While tedious, it is not too difficult to calculate all these. More realistic cases might involve, say, 15 objects, in which case the number of possible Z values is $15! \approx 1.3 \times 10^{12}$. Enumerating all of these then becomes impractical, and there are two possible approaches for carrying out the Mantel test. A large number of randomized E_R matrices can be generated on a computer and the resulting distribution of Z values used in place of the true randomized distribution. Alternatively, the mean, E(Z), and variance, Var(Z), of the randomized distribution of Z can be calculated, and

$$
g = \frac{Z - E(Z)}{Var(Z)}
$$

can be treated as a standard normal variate. Mantel (1967) provided formulae for the mean and variance of Z in the null hypothesis case of no correlation between the distance measures. There is, however, some doubt about the validity of the normal approximation for the test statistic g (Mielke, 1978); we recommend randomizations rather than this approximation.

The test statistic Z of Equation 5.5 is the sum of the products of the elements in the lower diagonal parts of the matrices $\mathbf{M}$ and $\mathbf{E}$. The only reason for using this particular statistic is that Mantel's equations for the mean and variance are available. However, when using randomizations, there is no reason why the test statistic should not be changed. Indeed, values of Z are not particularly informative except in comparison with the randomization mean and variance. It may, therefore, be more useful to take the correlation r_{ME} between the lower diagonal elements of $\mathbf{M}$ and $\mathbf{E}$ as the test statistic instead of Z. With $n \times n$ matrices, there are $n(n-1)/2$ lower diagonal terms, which pair up as $(m_{21}, e_{21}), (m_{31}, e_{31}), (m_{32}, e_{32})$, and so on. Their correlation is calculated in the usual way, as explained in Section 2.7.

The correlation r_{ME} has the usual interpretation in terms of the relationship between the two distance measures. Thus, r ranges from -1 to $+1$, with $r = -1$, indicating a perfect negative correlation, $r = 0$ indicating no correlation, and $r = +1$ indicating a perfect positive correlation. The significance or otherwise of the data will be the same for the test statistics Z and r because, in fact, there is a simple linear relationship between them.

5.6.1 Example 5.4: More on Distances between Samples of Egyptian Skulls

For the Egyptian skull data, we can ask whether the distances given in Table 5.4, based on four skull measurements, are significantly related to the time differences between the five samples. This certainly does seem to be the case, but a definitive answer is provided by Mantel's test.

The sample times are approximately 4000 BC (early predynastic), 3300 BC (late predynastic), 1850 BC (12th and 13th Dynasties), 200 BC (Ptolemaic), and AD 150 (Roman). Comparing Penrose's distance measures with time differences (in thousands of years), therefore, provides the lower diagonal distance matrices between the samples that are shown in Table 5.7. The correlation between the elements of these matrices is 0.954. It appears, therefore, that the distances agree very well.

TABLE 5.7

Penrose Distances Based on Skull Measurements and Time Differences (Thousands of Years) for Five Samples of Egyptian Skulls

Penrose distances					Time distances				
—					—				
0.023	—				0.70	—			
0.216	0.163	—			2.15	1.45	—		
0.493	0.404	0.108	—		3.80	3.10	1.65	—	
0.736	0.583	0.244	0.066	—	4.15	3.45	2.00	0.35	—

There are $5! = 120$ possible ways to reorder the five samples for one of the two matrices, and, consequently, there are 120 elements in the randomization distribution for the correlation. Of these, only one is larger than the observed correlation of 0.954. It follows that the observed correlation is significantly high at the $(2/120)100\% = 1.7\%$ level, and there is evidence of a relationship between the two distance matrices. A one-sided test is appropriate because there is no reason why the samples of skulls should become more similar as they get further apart in time.

The matrix correlation between Mahalanobis distances and time distance is 0.964. This is also significantly large at the 1.7% level when compared with the randomization distribution.

5.7 Computer Programs

Calculating distance and similarity measures is the first step in cluster analysis and ordination methods. This is often easiest to do using specially designed computer programs. However, the clustering and ordination options of more general statistical packages can be used. Mantel test on distance and similarity matrices can also be carried out using the R code described in the appendix to this chapter.

5.8 Discussion and Further Reading

The use of different measures of distance and similarity is the subject of continuing debates, indicating a lack of agreement about what is the best under different circumstances. The problem is that no measure is perfect, and the conclusions from an analysis may depend to some extent on which of several reasonable measures is used. The situation depends very much on the purpose of calculating the distances or similarities, and the nature of the data available.

The usefulness of the Mantel randomization method for testing for an association between two distance or similarity matrices has led to several proposals for methods to analyze relationships between three or more such matrices. These are reviewed by Manly and Navarro Alberto (2021). At present, a major unresolved problem in this area relates to the question of how to take proper account of the effects of spatial correlation when, as is often the case, the items between which distances and similarities are measured tend to be similar when they are relatively close in space.

Peres-Neto and Jackson (2001) have suggested that the comparison between two distance matrices using a method called Procrustes analysis is better than the use of the Mantel test. Procrustes analysis was developed by Gower

(1971, 2010) as a means of seeing how well two data configurations can be matched up after suitable manipulations. Peres-Neto and Jackson propose a randomization test to assess whether the matching that can be obtained with two distance matrices is significantly better than expected by chance. In contrast, a comparative study conducted by Forcino et al. (2015) based on distance matrices used in paleontological research concluded that the Mantel test and the Procrustes randomization test give similar results in most cases. Moreover, they showed that occasionally, the simpler Mantel test is better in properly detect correlations between distances matrices.

Exercise

Consider the data in Table 1.3.

1. Standardize the environmental variables altitude, annual precipitation, annual maximum temperature, and annual minimum temperature to means of zero and standard deviations of one, and calculate Euclidean distances between all pairs of colonies using Equation 5.1 to obtain an environmental distance matrix.

2. Use the *Pgi* gene frequencies, converted to proportions (p), to calculate genetic distances between colonies i and j as $d_{ij} = \sum |p_i - p_j| / 2$.

3. Carry out a Mantel matrix randomization test to determine whether there is a significant positive relationship between the environmental and genetic distances, and report your conclusions.

4. Explain why a significant positive relationship on a randomization test in a situation such as this could be the result of spatial correlations between the data for close colonies rather than from environmental effects on the genetic composition of colonies.

References

Batyrshin, I.Z., Kubysheva, N., Solovyev, V., and Villa-Vargas, L.A. (2016). Visualization of similarity measures for binary data and 2 × 2 tables. *Computación y Sistemas* 20(3): 345–353.

Brusco, M., Cradit, J.D., and Steinley, D. (2021). A comparison of 71 binary similarity coefficients: the effect of base rates. *PLoS ONE* 16(4): e0247751. https://doi.org/10.1371/journal.pone.0247751.

Cibulková, J., Šulc, Z., Řezanková, H., and Sirota, S. (2020). Associations among similarity and distance measures for binary data in cluster analysis. *Metodološki zvezki* 17(1): 33–54.

Forcino, F.L., Ritterbush, K.A., and Stafford, E.S. (2015). Evaluating the effectiveness of the Mantel test and Procrustes randomization test for exploratory ecological similarity among paleocommunities. *Palaeogeography, Palaeoclimatology, Palaeoecology* 426: 99–208.

Gower, J.C. (1971). Statistical methods for comparing different multivariate analyses of the same data. In Hodson, F.R., Kendall, D.G., and Tautu, P. (eds). *Mathematics in the Archaeological and Historical Sciences*, pp. 138–149. Edinburgh: Edinburgh University Press.

Gower, J.C. (2010). Procrustes methods. *WIREs Computational Statistics* 2: 503–508. https://doi.org/10.1002/wics.107.

Gower, J.C., and Legendre, P. (1986). Metric and non-metric properties of dissimilarity coefficients. *Journal of Classification* 5: 5–48.

Higham, C.F.W., Kijngam, A., and Manly, B.F.J. (1980). Analysis of prehistoric canid remains from Thailand. *Journal of Archaeological Science* 7: 149–165.

Jackson, D.A., Somers, K.M., and Harvey, H.H. (1989). Similarity coefficients: measures of co-occurrence and association or simply measures of co-occurrence. *American Naturalist* 133: 436–453.

Mahalanobis, P.C. (1948). Historic note on the D^2-statistic. *Sankhya* 9: 237.

Manly, B.F.J., and Navarro Alberto, J.A. (2021). *Randomization, Bootstrap and Monte Carlo Methods in Biology*. 4th Edn. Boca Raton, FL: CRC Press.

Mantel, N. (1967). The detection of disease clustering and a generalized regression approach. *Cancer Research* 27: 209–220.

Mielke, P.W. (1978). Classification and appropriate inferences for Mantel and Varland's nonparametric multivariate analysis technique. *Biometrics* 34: 272–282.

Penrose, L.W. (1953). Distance, size and shape. *Annals of Eugenics* 18: 337–343.

Peres-Neto, P.R., and Jackson, D.A. (2001). How well do multivariate data sets match? The advantages of a Procrustean superimposition approach over a Mantel test. *Oecologia* 129: 169–178.

Appendix: Multivariate Distance/ Similarity Measures in R

A.1 Calculation of Distance and Similarity Measures

A.1.1 Distances between Individual Observations

The main function in R for distance calculations is `dist`, from which the user can choose among six distance measures, the default being the Euclidean distance. The most usual arguments for `dist` are either a numeric matrix or a data frame, and distances are computed between pairs of rows. If distances between columns are sought, the transpose operator `t()`, described in the appendix for Chapter 2, must be applied. Additional formatting options are available for the output of the computed distance matrix—for example, showing only the upper or lower triangular matrix.

The R script needed to obtain the Euclidean distances between dogs and related species (Example 5.1) is provided in the supplementary material at the book's website. According to the procedure seen in that example, first, the variables are standardized using `scale` or with the function `decostand` found in the package `vegan` (Oksanen et al., 2024), and then `dist` is applied to the standardized data. `scale` has a simpler syntax than `decostand` and this latter requires the numeric (column) data vectors to be put into matrix form. In fact, `vegan` provides two other functions (`vegdist` and `designdist`) with further distance measures, some of them special for binary data, as described in the following. An additional package also offering different distance measures is `ecodist` (Goslee and Urban, 2007), especially created for R users working with ecological data. The main function in `ecodist` is `distance`, with seven dissimilarity metrics to choose, depending on the data's scale of measurement.

A.1.2 Distances between Populations and Samples

For the calculation of the Penrose distance, no special R functions or packages are directly available. To solve this limitation, the function `Penrose.dist` has been implemented in the source file **Penrose.dist.R** and made available at the repository of supplementary resources. `Penrose.dist` has two arguments: x, a data frame with $p + 1$ columns containing p numeric variables, and `group`, the factor that classifies samples; the function requires package `biotools` (da Silva, 2017, 2021), described in the appendix to Chapter 4. Moreover, `biotools` provides an specific R function for the computation of Mahalanobis distances with function `D2.dist`, while the function `distance` implemented in `ecodist` allows the specification of the Mahalanobis distance as one of its options. R scripts for producing the Penrose and Mahalanobis distances between pairs of samples of Egyptian skulls, as in Table 5.4, can be found at the book's website. These scripts take advantage of the immediate availability of the estimated pooled covariance matrix produced by function `BoxM` from the package `biotools`.

A.1.3 Distances and Similarities for Presence-Absence Data

In the case of multivariate binary distances in R, the user can make use of the options available in different packages, starting from the default package, `stats`, and some others, such as `ade4` (Dray and Dufour, 2007), `vegan` (Oksanen et al., 2024), `labdsv` (Roberts, 2023) and `ecodist` (Goslee and Urban, 2007). Thus, the function `dist` from the `stats` package allows the computation of the Jaccard distance:

```
dist(data, method = "binary")
```

Here, `data` is a data object (matrix or data frame) containing either binary or nonnegative numbers, such that nonzero elements are converted into ones, and zero elements are kept intact. Jaccard's distances are also available in `vegan` (Oksanen et al., 2024) as one of the options of function `vegdist`; in `ade4` (Dray and Dufour, 2007) as one of the options of function `dist.binary`; in `labdsv` package (Roberts, 2023), with function `dsvdis`, option `index = "steinhaus"`, which is another name given to Jaccard's index; and in `ecodist` package (Goslee and Urban, 2007), with function `distance`, option `method = "jaccard"`. If Jaccard's similarity or any similarity index is desired, these are calculated as $s = 1 - d$, where d is a distance measure (e.g., Jaccard) and s is the corresponding similarity index. This relation applies for distance functions implemented in `stats`, `vegan` and `labdsv`. In the case of `ade4`, the ten different distances offered for binary data are of the form $d = \sqrt{1-s}$, where s is the associated similarity coefficient. Therefore, to compute a similarity index in `ade4`, the user must first invoke the corresponding distance d and then compute the similarity index as $s = 1 - d^2$. The simple matching coefficient is one of the distance measures included in `ade4`. Finally, both Ochiai and Dice-Sorensen similarity measures can be calculated from the corresponding dissimilarities offered by functions `dist.binary` and `svdis` from packages `ade4` and `labdsv` respectively, while the `distance` function implemented in `ecodist` offers the Dice-Sorensen index but not Ochiai's. The supplementary material exemplifies the calculation of the four similarity measures listed in Section 5.5, in the analysis of the toy data given in Table 5.5, using the R-functions described earlier.

A.2 The Mantel Randomization Test

The package `vegan` (Oksanen et al., 2024) provides the function `mantel` for the calculation of the Mantel statistic and its corresponding randomization test of significance. Given two distance matrices `xdis` and `ydis`, the simplest expression for invoking the `mantel` function is

```
library(vegan)

mantel(xdis, ydis)
```

The user can include the argument `permutations =` if the number of elements in the randomization distribution for the correlation is very large. This is not the case for Example 5.3, where the number of all possible permutations is only 5! = 120. The Mantel test can also be performed over all the permutations with the `mantel.test` function, available from the `cultevo` package (Stadler, 2018), if the number of permutations is not too large. In contrast, the following packages only allow sampling of the randomization distribution of correlations: the `ade4` package (Dray and Dufour, 2007; Thioulouse et al., 2018) via the function `mantel.rtest`; the package `biotools` (da Silva, 2017, 2021) given by function `mantelTest`; the `ecodist` package (Goslee and Urban, 2007), with `mantel`, the same function name given by `vegan`; and the `ape` package (Paradis and Schliep, 2019), with function `mantel.test`, the same function name assigned by the `cultevo` package. As a way to refer to a package function when more than one package is loaded and they contain the same function name, it is recommended to use the double colon "operator" `::`. As an example, if you want to run `ecodist`'s function `mantel` with the default number of permutations (`nperm = 1000`) for Mantel's test, you can write

```
ecodist :: mantel(xdis ~ ydis)
```

A script containing all the functions listed has been created and applied in the analysis of Egyptian skulls data (Example 5.3), and this script can be found at this book's website.

References

da Silva, A.R. (2021). *biotools: Tools for Biometry and Applied Statistics in Agricultural Science*. R package version 4.2. https://cran.r-project.org/package=biotools.
da Silva, A.R., Malafaia, G., and Menezes, I.P.P. (2017). biotools: an R function to predict spatial gene diversity via an individual-based approach. *Genetics and Molecular Research* 16. https://doi.org/10.4238/gmr16029655.

Dray, S., and Dufour, A.B. (2007). The ade4 package: implementing the duality diagram for ecologists. *Journal of Statistical Software* 22(4): 1–20.

Goslee, S.C., and Urban, D.L. (2007). The ecodist package for dissimilarity-based analysis of ecological data. *Journal of Statistical Software* 22(7): 1–19. https://doi.org/10.18637/jss.v022.i07.

Oksanen, J., Simpson, G., Blanchet, F., Kindt, R., Legendre, P., Minchin, P., O'Hara, R., Solymos, P., Stevens, M., Szoecs, E., Wagner, H., Barbour, M., Bedward, M., Bolker, B., Borcard, D., Carvalho, G., Chirico, M., De Caceres, M., Durand, S., Evangelista, H., FitzJohn, R., Friendly, M., Furneaux, B., Hannigan, G., Hill, M., Lahti, L., McGlinn, D., Ouellette, M., Ribeiro Cunha, E., Smith, T., Stier, A., Ter Braak, C., and Weedon, J. (2024). *vegan: Community Ecology Package*. R package version 2.6-6.1. https://CRAN.R-project.org/package=vegan.

Paradis, E., and Schliep, K. (2019). ape 5.0: an environment for modern phylogenetics and evolutionary analyses in R. *Bioinformatics* 35: 526–528.

Roberts, D.W. (2023). *labdsv: Ordination and Multivariate Analysis for Ecology*. R package version 2.1-0. https://CRAN.R-project.org/package=labdsv.

Stadler, K. (2018). *cultevo: Tools, Measures and Statistical Tests for Cultural Evolution*. R package version 1.0.2. https://kevinstadler.github.io/cultevo/.

Thioulouse, J., Dray, S., Dufour, A., Siberchicot, A., Jombart, T., and Pavoine, S. (2018). *Multivariate Analysis of Ecological Data with ade4*. Springer. https://doi.org/10.1007/978-1-4939-8850-1.

6

Principal Components Analysis

6.1 Definition of Principal Components

The technique of principal components analysis was first described by Karl Pearson (1901). He apparently believed that this was the correct solution to some of the problems that were of interest to biometricians at that time, although he did not propose a practical method of calculation for more than two or three variables. A description of practical computing methods came much later from Hotelling (1933). Even then, the calculations were extremely daunting for more than a few variables, because they had to be done by hand. It was not until computers became generally available that the technique achieved widespread use.

Principal components analysis is one of the simplest of the multivariate methods. The object of the analysis is to take p variables $X_1, X_2, \ldots, X_p$ and find combinations of these to produce indices $Z_1, Z_2, \ldots, Z_p$ that are uncorrelated and in order of their importance in terms of the variation in the data. The lack of correlation means that the indices are measuring different dimensions of the data, and the ordering is such that $\mathrm{Var}(Z_1) \geq \mathrm{Var}(Z_2) \geq \ldots \geq \mathrm{Var}(Z_p)$, where $\mathrm{Var}(Z_i)$ denotes the variance of Z_i. The Z indices are then the principal components. When doing a principal components analysis, there is always the hope that the variances of most of the indices will be so low as to be negligible. In that case, most of the variation in the full data set can be adequately described by the few Z variables with variances that are not negligible, and some degree of economy is then achieved.

The foregoing might appear opaque and unduly abstract. Let us step back and look at a conceptual visualization of principal components for two variables; hopefully this visualization will help in understanding. Figure 6.1 is a scatterplot of two variables, Y_2 versus Y_1. We conventionally locate points by noting their distance perpendicular to the axes. For instance, the point in the lower left is $(Y_1, Y_2) = (63.4, 97.9)$. To locate that point, we suppose you automatically looked up from the horizontal axis (at 63.4) and over from the vertical axis (at 97.9).

It is conventional in multivariate analyses to first standardize the variables; here that leads to $X_{i1} = (Y_{i1} - \bar{Y}_1)/S_{Y_1}$; $i = 1, 2, \ldots, n$ and

DOI: 10.1201/9781003453482-6

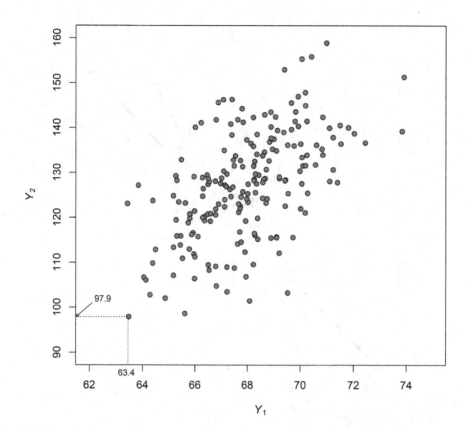

FIGURE 6.1
Scatterplot of two variables Y_1 and Y_2.

$X_{i2} = (Y_{i2} - \bar{Y}_2)/S_{Y_2}$; $i = 1, 2, ..., n$,, where $\bar{Y}_1$ and $\bar{Y}_2$ and S_{Y_1} and S_{Y_2} are the respective sample means and standard deviations. Thus, each variable now has a mean of zero and standard deviation of one, as depicted in Figure 6.2. This has the effect that they each have equal weight going into the analysis. It's worth noting that the numerical value of the mean and SD of a variable is a function of our choices for measurement units; standardizing takes away the impact of those arbitrary choices.

This rescaling reminds us that our choice of axes is to some extent arbitrary, chosen for one purpose or another. There are three lines drawn on this graph, A, B and C; one of them is the regression of X_2 on X_1. Upon seeing this sort of plot for the first time, most beginners would choose Line B as this regression line. Recall that a regression line will go through the point determined by the mean of each variable: $(\bar{X}_1, \bar{X}_2) = (0, 0)$ here. All three lines meet that criterion but the regression of X_2 on X_1 is *not* Line B but Line C. The purpose of a regression model is to estimate the mean of the response (X_2) here, given a value for the predictor (X_1). Pick a value (or three) for X_1;

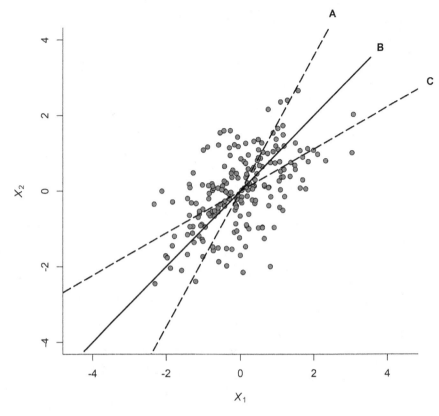

FIGURE 6.2
Graphical interpretation of principal component analysis in two dimensions for the standardization (X_1 and X_2) of the data (Y_1 and Y_2) depicted in Figure 6.1. Line A is the regression of X_1 on X_2, and C is the regression line of X_2 on X_1. Line B is the first principal component for X_1 and X_2.

overall, Line C does the best job of hitting the center of the values for X_2. In fact, Line A is the regression of X_1 on X_2 (the orientation of the graph dictates to choose some values for X_2, then look horizontally, and it will be seen that Line A, overall, does the job of predicting the mean of X_1).

What, then, is Line B? It is the solution to the following challenge: draw a line on the graph that gets as close as possible to all the points ("close" being measured perpendicular to said line). It is in fact a depiction of the first principal component, which can be thought of as another choice of an axis for this graph. You can imagine that this line would take the form $Z_1 = c_1 X_1 + c_2 X_2$, where the coefficients c_1 and c_2 are each positive numbers less than one. In the spirit of how we usually draw axes, a second axis would go through $(0,0)$ and itself be perpendicular to the first axis. In fact, this line (not shown, to reduce chart-clutter) would depict the second principal component, expressed as $Z_2 = b_1 X_1 - b_2 X_2$, both b_1 and b_2 being positive and less than one.

Typically, the first principal component is indeed a weighted sum of the variables (we cannot call it a "weighted average" because for that the coefficients need to add to one, which they typically do not here). And the second is usually the difference between one collection of variables and another. Sometimes this feature lends itself to naming the components in a biologically useful way. For instance, suppose the original data here were height and weight. The first principal component could be called "size" and the second "shape."

Principal components analysis and any other multivariate method are worth applying when there are quite more than only two variables (but two served well for our graphs presented). If you begin with five or six variables and decide that two principal components are satisfactory, then you will have reduced the dimensionality (i.e., complexity) of the data considerably. However, principal components analysis does not always work; indeed, if the original variables are uncorrelated, then the analysis achieves nothing. The best results are obtained when the original variables are very highly correlated, positively or negatively. If that is the case, then as many as 20 or more original variables can be adequately represented by two or three principal components. In that case, the important principal components will be of some interest as measures of the underlying dimensions in the data. It will also be of value to know that there is a good deal of redundancy in the original variables, with most of them measuring similar things.

Before describing the calculations involved in a principal components analysis, we will look briefly at the outcome of the analysis when it is applied to the data in Table 1.1 on five body measurements of 49 female sparrows. Details are given in Example 6.1. Here, the five measurements are quite highly correlated, as shown in Table 6.1. The first principal component has a variance of 3.62, whereas the other components all have variances that

TABLE 6.1

Correlations between the Five Body Measurements of Female Sparrows Calculated from the Data of Table 1.1

	X_1	X_2	X_3	X_4	X_5
X_1, total length	1.000	–			
X_2, alar extent	0.735	1.000	–		
X_3, length of beak and head	0.662	0.674	1.000	–	
X_4, length of humerus	0.645	0.769	0.763	1.000	–
X_5, length of kneel of sternum	0.605	0.529	0.526	0.607	1.000

Note: Only the lower part of the table is shown, because the correlation between X_i and X_j is the same as the correlation between X_j and X_i.

are much less than this (0.53, 0.39, 0.30, and 0.16). This means that the first principal component is by far the most important of the five components for representing the variation in the measurements of the 49 birds. The first component is calculated to be

$$Z_1 = 0.45X_1 + 0.46X_2 + 0.45X_3 + 0.47X_4 + 0.40X_5$$

where X_1–X_5 denote the measurements in Table 1.1 in order, after they have been standardized to have zero means and unit standard deviations.

Clearly, Z_1 behaves like an average of the standardized body measurements, and it can be thought of as a simple index of size. The analysis, therefore, leads to the conclusion that most of the differences between the 49 birds are a matter of size rather than shape.

6.2 Procedure for a Principal Components Analysis

A principal components analysis starts with data on p variables for n individuals, as indicated in Table 6.2.

The first principal component is then the linear combination of the variables $X_1, X_2, \ldots, X_p$:

$$Z_1 = a_{11}X_1 + a_{12}X_2 + \cdots + a_{1p}X_p$$

that varies as much as possible for the individuals, subject to the condition that

$$a_{11}^2 + a_{12}^2 + \cdots + a_{1p}^2 = 1.$$

TABLE 6.2

The Form of Data for a Principal Components Analysis, with Variables X_1 to X_p and Observations on n Cases

Case	X_1	X_2	$\cdots$	X_p
1	x_{11}	x_{12}	$\cdots$	x_{1p}
2	x_{21}	x_{22}	$\cdots$	x_{2p}
$\vdots$	$\vdots$	$\vdots$	$\ddots$	$\vdots$
n	x_{n1}	x_{n2}	$\cdots$	x_{np}

Thus, $\text{Var}(Z_1)$, the variance of Z_1, is as large as possible given this constraint on the constants a_{1j}. The constraint is introduced because if this is not done, then $\text{Var}(Z_1)$ can be increased by simply increasing any one of the a_{1j} values.

The second principal component,

$$Z_2 = a_{21}X_1 + a_{22}X_2 + \cdots + a_{2p}X_p,$$

is chosen so that $\text{Var}(Z_2)$ is as large as possible, subject to the constraint that

$$a_{21}^2 + a_{22}^2 + \cdots + a_{2p}^2 = 1,$$

and also to the condition that Z_1 and Z_2 have zero correlation for the data.

The third principal component,

$$Z_3 = a_{31}X_1 + a_{32}X_2 + \cdots + a_{3p}X_p,$$

is such that $\text{Var}(Z_3)$ is as large as possible, subject to the constraint that

$$a_{31}^2 + a_{32}^2 + \cdots + a_{3p}^2 = 1,$$

and also that Z_3 is uncorrelated with both Z_1 and Z_2. Further principal components are defined by continuing in the same way. If there are p variables, then there will be up to p principal components.

To use the results of a principal components analysis, it is not necessary to know how the equations for the principal components are derived. However, it is useful to understand the nature of the equations themselves. In fact, a principal components analysis involves finding the eigenvalues of the sample covariance matrix.

The calculation of the sample covariance matrix has been described in Sections 2.6 and 2.7. The covariance matrix is symmetric and has the form

$$\mathbf{C} = \begin{bmatrix} c_{11} & c_{12} & \cdots & c_{1p} \\ c_{21} & c_{22} & \cdots & c_{2p} \\ \vdots & \vdots & \ddots & \vdots \\ c_{p1} & c_{p2} & \cdots & c_{pp} \end{bmatrix} \tag{6.1}$$

where the diagonal element c_{pp} is the variance of X_i, and the off-diagonal terms $c_{ij} = c_{ji}$ are the covariance of variables X_i and X_j.

The variances of the principal components are the eigenvalues of the matrix $\mathbf{C}$. There are p of these eigenvalues, some of which may be zero, but negative eigenvalues are not possible for a covariance matrix. Assuming that the eigenvalues are ordered as $\lambda_1 \geq \lambda_2 \geq ... \geq \lambda_p$, then λ_i corresponds to the ith principal component

$$Z_i = a_{i1}X_1 + a_{i2}X_2 + \cdots + a_{ip}X_p$$

In particular, $\text{Var}(Z_i) = \lambda_i$ and the constants $a_{i1}, a_{i2}, \ldots, a_{ip}$ are the elements of the corresponding eigenvector, scaled so that

$$a_{i1}^2 + a_{i2}^2 + \cdots + a_{ip}^2 = 1.$$

An important property of the eigenvalues is that they add up to the sum of the diagonal elements (the trace) of the matrix $\mathbf{C}$. That is,

$$\lambda_1 + \lambda_2 + \cdots + \lambda_p = c_{11} + c_{22} + \cdots + c_{pp}.$$

As c_{ii} is the variance of X_i and λ_i is the variance of Z_i, this means that the sum of the variances of the principal components is equal to the sum of the variances of the original variables. Therefore, in a sense, the principal components account for all the variation in the original data. Given that the variables have been standardized, the matrix $\mathbf{C}$ takes the form

$$\mathbf{C} = \begin{bmatrix} 1 & c_{12} & \cdots & c_{1p} \\ c_{21} & 1 & \cdots & c_{2p} \\ \vdots & \vdots & \ddots & \vdots \\ c_{p1} & c_{p2} & \cdots & 1 \end{bmatrix}$$

where $c_{ij} = c_{ji}$ is the correlation between X_i and X_j. In other words, the principal components analysis is carried out on the correlation matrix. In that case, the sum of the diagonal terms, and hence the sum of the eigenvalues, is equal to p, the number of X variables.

The steps in a principal components analysis are as follows:

1. Code the variables $X_1, X_2, \ldots, X_p$, to have zero means and unit variances. This is omitted in cases where the importance of variables is reflected in their variances.

2. Calculate the covariance matrix $\mathbf{C}$. This is a correlation matrix if Step 1 has been done.

3. Find the eigenvalues $\lambda_1, \lambda_2, \ldots, \lambda_p$ and the corresponding eigenvectors $\mathbf{a}_1, \mathbf{a}_2, \ldots, \mathbf{a}_p$. The coefficients of the ith principal component are the elements of $\mathbf{a}_i$, while λ_i is its variance.

4. Discard any components that only account for a small proportion of the variation in the data. For example, starting with 20 variables, it may be that the first three components account for 90% of the total variance. Then the other 17 components may reasonably be ignored.

6.2.1 Example 6.1: Body Measurements of Female Sparrows

Some mention has already been made of what happens when a principal components analysis is carried out on the data on five body measurements of 49 female sparrows (Table 1.1). This example is now considered in more detail.

It is appropriate to begin with Step 1 of the four parts of the analysis that have just been described. Omitting standardization would mean that the variables X_1 and X_2, which vary most over the 49 birds, would tend to dominate the principal components.

The covariance matrix for the standardized variables is the correlation matrix. This has already been given in lower triangular form in Table 6.1. The eigenvalues of this matrix are found to be 3.616, 0.532, 0.386, 0.302, and 0.165. These add to 5.000, the sum of the diagonal terms in the correlation matrix. The corresponding eigenvectors are shown in Table 6.3, standardized so that the sum of the squares of the coefficients is one for each of them. These eigenvectors then provide the coefficients of the principal components.

The eigenvalue for a principal component indicates the variance that it accounts for out of the total variances of 5.000. Thus, the first principal component accounts for (3.616/5.000)100% = 72.3% of the total variance. Similarly, the other principal components in order account for 10.6%, 7.7%, 6.0%, and 3.3%, respectively, of the total variance. Clearly, the first component is far more important than any of the others.

TABLE 6.3

The Eigenvalues and Eigenvectors of the Correlation Matrix for Five Measurements on 49 Female Sparrows

		Eigenvectors (coefficients for the principal components)				
Component	Eigenvalue	X_1	X_2	X_3	X_4	X_5
1	3.616	0.452	0.462	0.451	0.471	0.398
2	0.532	−0.051	0.300	0.325	0.185	−0.876
3	0.386	−0.690	−0.341	0.454	0.411	0.178
4	0.302	0.420	−0.548	0.606	−0.388	−0.069
5	0.165	0.374	−0.530	−0.343	0.652	−0.192

Note: The eigenvalues are the variances of the principal components. The eigenvectors give the coefficients of the standardized X variables used to calculate the principal components.

Another way of looking at the relative importance of principal components is in terms of their variance as compared to that of the original variables. After standardization, the original variables all have variances of 1.0. The first principal component, therefore, has a variance of 3.62 times that of the original variables. However, the second principal component has a variance of only 0.532 of that of one of the original variables, while the other principal components account for even less. This confirms the importance of the first principal component in comparison with the others.

The first principal component is

$$Z_1 = 0.452X_1 + 0.462X_2 + 0.451X_3 + 0.471X_4 + 0.398X_5$$

where X_1 to X_5. The coefficients of the X variables are nearly equal, and this is clearly an index of the size of the sparrows. It seems, therefore, that about 72.3% of the variation in the data is related to size differences among the sparrows.

The second principal component is

$$Z_2 = -0.051X_1 + 0.300X_2 + 0.325X_3 + 0.185X_4 - 0.876X_5.$$

This is a contrast between variables X_2 (alar extent), X_3 (length of beak and head), and X_4 (length of humerus), on the one hand, and variable X_5 (length of the keel of the sternum), on the other. That is to say, Z_2 will be high if X_2, X_3, and X_4 are high but X_5 is low. On the other hand, Z_2 will be low if X_2, X_3, and X_4 are low but X_5 is high. Hence, Z_2 represents a shape difference between the sparrows. The low coefficient of X_1 (total length) means that the value of this variable does not affect Z_2 very much. The other principal components can be interpreted in a similar way. They, therefore, represent other aspects of shape differences.

The values of the principal components may be useful for further analyses. They are calculated in the obvious way from the standardized variables. Thus, for the first bird, the original variable values are $x_1 = 156$, $x_2 = 245$, $x_3 = 31.6$, $x_4 = 18.5$, and $x_5 = 20.5$. These standardize to $x_1 = (156 - 157.980)/3.654 = -0.542$, $x_2 = (245 - 241.327)/5.068 = 0.725$, $x_3 = (31.6 - 31.459)/0.795 = 0.177$, $x_4 = (18.5 - 18.469)/0.564 = 0.055$, and $x_5 = (20.5 - 20.827)/0.991 = -0.330$, where in each case the variable mean for the 49 birds has been subtracted, and a division has been made by the sample standard deviation for the 49 birds. The value of the first principal component for the first bird is, therefore,

$$Z_1 = 0.452 \times (-0.542) + 0.462 \times 0.725 + 0.451 \times 0.177$$
$$+ 0.471 \times 0.055 + 0.398 \times (-0.330)$$
$$= 0.064.$$

The second principal component for the same bird is

$$Z_2 = -0.051 \times (-0.542) + 0.300 \times 0.725 + 0.325 \times 0.177$$
$$+ 0.185 \times 0.055 - 0.877 \times (-0.330)$$
$$= 0.602.$$

The other principal components can be calculated in a similar way.

The birds being considered were picked up after a severe storm; 21 of them recovered, while the other 28 died. A question of some interest is, therefore, whether the survivors and nonsurvivors show any differences. It has been shown in Example 4.1 that there is no evidence of any differences in mean values. However, in Example 4.2, it has been shown that the survivors seem to have been less variable than the nonsurvivors. The situation will now be considered in terms of principal components.

The means and standard deviations of the five principal components are shown in Table 6.4 separately for survivors and nonsurvivors. None of the mean differences between survivors and nonsurvivors are significant from t-tests, and none of the standard deviation differences are significant on F-tests. However, Levene's test on deviations from medians (described in Chapter 4) just gives a significant difference between the variation of principal component 1 for survivors and nonsurvivors on a one-sided test at the 5% level. The assumption for the one-sided test is that, if anything, nonsurvivors were more variable than survivors. The variation is not significantly different for survivors and nonsurvivors with Levene's test on the other principal components. As principal component 1 measures overall size, it seems that stabilizing selection may have acted against very large and very small birds.

Figure 6.3 shows a plot of the values of the 49 birds for the first two principal components, which between them account for 82.9% of the variation in the data. The figure shows quite clearly how birds with extreme values for the first principal component failed to survive. Indeed, there is a suggestion that this was true for principal component 2 as well.

It is important to realize that some computer programs may give the principal components as shown with this example but with the signs of the coefficients of the body measurements reversed. For example, Z_2 might be shown as

$$Z_2 = 0.051X_1 - 0.300X_2 - 0.325X_3 - 0.185X_4 + 0.876X_5.$$

This is not a mistake. The principal component is still measuring exactly the same aspect of the data, but in the opposite direction.

TABLE 6.4

Comparison between Survivors and Nonsurvivors in Terms of Means and Standard Deviations of Principal Components

Principal Component	Mean		Standard deviation	
	Survivors	Nonsurvivors	Survivors	Nonsurvivors
1	–0.100	0.075	1.506	2.176
2	0.004	–0.003	0.680	0.776
3	0.140	–0.105	0.522	0.677
4	–0.073	0.055	0.563	0.543
5	0.023	–0.017	0.411	0.408

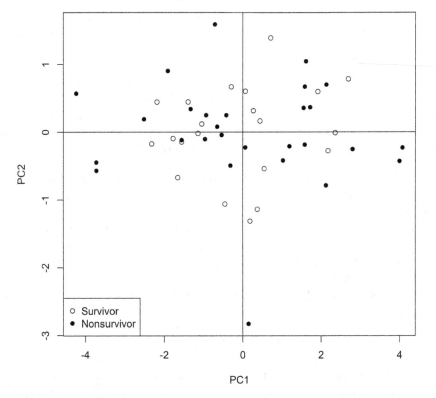

FIGURE 6.3

Plot of 49 female sparrows against values for the first two principal components, PC1 and PC2.

6.2.2 Example 6.2: Cost of a Healthy Diet in Europe

As a second example of a principal components analysis, consider the data (FAO, 2023) presented in Table 1.5 on the cost of a healthy diet in European countries. The correlation matrix for the six variables is shown in Table 6.5. Overall, some values in this matrix are not particularly high, which indicates

TABLE 6.5

The Correlation Matrix of the Cost of Six Food Groups for 38 European Countries in Lower Diagonal Form, Calculated from the Data in Table 1.5.

	X_1	X_2	X_3	X_4	X_5	X_6
X_1	1.000	—				
X_2	0.723	1.000	—			
X_3	0.559	0.498	1.000	—		
X_4	0.246	0.141	−0.015	1.000	—	
X_5	0.377	0.568	0.434	0.346	1.000	—
X_6	0.626	0.552	0.468	−0.144	0.057	1.000

Note: The variables are the costs (purchasing power parity dollar per person per day, or PPP dollar per person per day) of individual food groups that make up a healthy diet basket: X_1 = CSTS, starchy staples; X_2 = CASF, animal source foods; X_3 = CLNS, legumes, nuts, and seeds; X_4 = CVEG; vegetables; X_5 = CFRT, fruits; X_6 = COFT, oils and fats.

that more than one principal component will be required to account for the variation in the data.

The eigenvalues of the correlation matrix, with percentages of the total of 6.000 in parentheses, are 3.023 (50.4%), 1.322 (22.0%), 0.729 (12.1%), 0.486 (8.1%), 0.258 (4.3%), and 0.183 (3.0%). The last eigenvalue has the value 0.183 for all the countries, and hence, the last principal component has very low variance. If any linear combination of the original variables in a principal components analysis is constant or nearly constant (i.e., it has zero or very low variance), then this must of necessity result in one of the eigenvalues being zero or close to zero.

This example is not as straightforward as the previous one. The first principal component only accounts for about 50% of the variation in the data, and three components are needed to account for about 85% of the variation. It is a matter of judgment as to how many components are important. It can be argued that only the first two should be considered, because these are the ones with eigenvalues greater than one. To some extent, the choice of the number of components that are important will depend on the use that is going to be made of them. For the present example, it will be assumed that a small number of indices are required to present the main aspects of differences between the countries, and for simplicity, only the first two components will be examined further. Between them, they account for about 72% of the variation in the original data.

The first component is

$$Z_1 = -0.505(\text{CSTS}) - 0.505(\text{CASF}) - 0.432(\text{CLNS}) - 0.127(\text{CVEG})$$
$$- 0.362(\text{CFRT}) - 0.395(\text{COFT}) \tag{6.2}$$

where the abbreviations for variables are defined in Table 6.5. As the analysis has been done on the correlation matrix, the variables in this equation are the purchasing power parity dollar per person per day (PPP) after they have each been standardized to have a mean of zero and a standard deviation of one. From the coefficients of Z_1, it can be seen that they are all negative. The largest contributions (largest absolute values) to the component are dictated by the costs of food groups not coming from vegetables, but from the other groups, with starchy staples (CSTS) and animal source foods (CASF) having the most prominent role and the smallest role is played by the fruit group (CFRT).

The second component is

$$Z_2 = -0.041(\text{CSTS}) + 0.017(\text{CASF}) - 0.139(\text{CLNS}) + 0.716(\text{CVEG}) \\ + 0.479(\text{CFRT}) - 0.487(\text{COFT}) \tag{6.3}$$

which primarily contrasts the vegetable (CVEG) and fruit (CFRT) groups to the oils and fat group (COFT). The contributions of the other food groups are insignificant.

Figure 6.4 shows a plot of the 38 countries against their values for Z_1 and Z_2. The picture is certainly rather meaningful in terms of what is known about

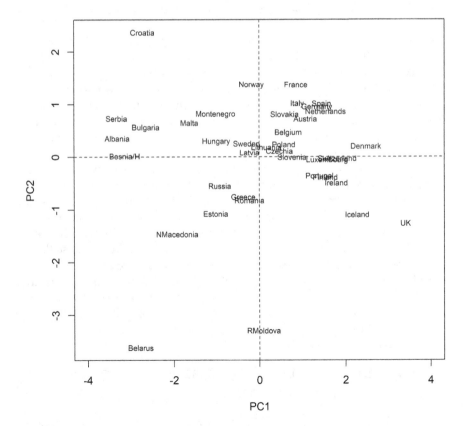

FIGURE 6.4
European countries plotted against the first two principal components for variables indicating the cost of individual food groups making up a healthy diet basket.

states. Most of the Balkan and Eastern European countries are grouped with negative values for Z_1 and values for Z_2 between about plus minus two. Most of the traditional Western democracies are grouped with positive values for Z_1, with the United Kingdom having the largest value. Belarus and the Republic of Moldova stand out as having rather distinct purchasing power of healthy food per person per day, while the remaining countries lie in a band ranging from Croatia ($Z_1 = -2.7$, $Z_2 = 2.4$) to North Macedonia ($Z_1 = -1.9$, $Z_2 = -1.5$).

As with the previous example, it is possible that some computer programs will produce the principal components shown here, but with the signs of the coefficients of the original variables reversed. The components still measure the same aspects of the data, but with the high and low values reversed.

6.3 Principal Components Analysis and Missing Data

How to deal with missing data for multivariate analyses has already described in Chapter 2. A solution for principal component analysis not involving data imputation of the missing observations, is to compute correlations or covariances using pairwise-complete observations. If principal components are obtained from the correlation matrix, Dray and Josse (2015) give the acronym *PairCor* for this strategy, and Podani et al. (2021) suggest a natural extension of this strategy, called *PairCov*, for the covariance matrix. The four steps for the principal components analysis listed in Section 6.2 (assuming no-missing data) are basically the same for those using the PairCor/PairCov strategies for missing data. The only difference resides in the second step, where the C matrix is computed using only the observed values for each pair of variables independently. This approach has the restriction of constraining each pair of variables to have data available for at least two common sampling units. In addition, with PairCor or PairCov, different correlation/covariance coefficients are not necessarily based on the same individuals or the same number of individuals. van Ginkel (2023) points out that although the pairwise-complete observations method uses more information from the data than the complete-case observations method does (described in Section 6.4), an implicit assumption is that the missingness mechanism for the data must be missing completely at random, i.e., there is no relation between the missing values and any observed or unobserved variable. An additional disadvantage of pairwise deletion is that combinations of correlations may yield a correlation matrix that is not positive semi-definite, giving place to a decomposition of eigenvalues with the possibility to produce negative eigenvalues. Dray and Josse (2015) suggest that the associated dimensions for components with negative eigenvalues should not be interpreted.

6.3.1 Example 6.3 Principal Component Analysis of the Costs of a Healthy Diet in Europe with Missing Data

The application of principal component analysis is illustrated with basically the same data given in Table 1.5 and Example 6.2, corresponding the costs of purchasing six food groups in 38 European countries, but individual values of the variables listed in Table 6.6 are assumed to be missing at random:

TABLE 6.6

Sample Sizes (n_i) of Variables Presented in Example 6.2, after Individual Observations of the Countries Listed Became Missing

Variable	Country with missing values	n_i
X_1	Belgium	37
X_2	Luxembourg, Netherlands, Sweden	35
X_3	Greece, Italy, Switzerland	35
X_4	Latvia, Switzerland	36
X_5	—	38
X_6	—	38
	Missing observations = 9	

Table 6.7 displays the correlations between the six variables with nine missing values, using the pairwise-complete observations method, and the corresponding sample sizes used for the calculation of those correlations.

TABLE 6.7

The Correlation Matrix of the Costs of Purchasing Six Food Groups in 38 European Countries with Missing Data, Computed Using the Pairwise-Complete Observations Method. The "effective" sample size is shown in parenthesis, below each corresponding correlation.

	X_1	X_2	X_3	X_4	X_5	X_6
X_1	1.000 (37)					
X_2	0.765 (34)	1.000 (35)				
X_3	0.543 (34)	0.486 (32)	1.000 (35)			
X_4	0.259 (35)	0.150 (33)	0.001 (34)	1.000 (36)		
X_5	0.402 (37)	0.580 (35)	0.461 (35)	0.376 (36)	1.000 (38)	
X_6	0.617 (37)	0.539 (35)	0.447 (35)	−0.178 (36)	0.057 (38)	1.000 (38)

The correlation between X_2 and X_3 have been calculated using an "effective" sample size of 32 observations only, given that the three missing observations in X_2 are different from the single missing observation in X_3, leaving $38 - (3 + 3) = 32$ pairs of complete observations. Also, not all the 34 pairs of complete observations used to compute the correlation between X_1 and X_3 are the same as those used to compute the correlation between X_1 and X_4. These instances illustrate the principle behind the pairwise-complete observations method, being all other cases similar. The correlation matrices for the complete (Table 6.5) and incomplete health data (Table 6.7) are not remarkably different, but it cannot be assured that this will happen for other data sets using the pairwise-complete observations method, or any other method chosen to handle missing data (e.g., data imputation). The real fact is that once data are missing, it is almost impossible to compare results with non-existing complete data!

Completing the last two steps of the principal component analysis listed in Section 6.2, it is found that the eigenvalues of the correlation matrix, with percentages of the total of 6.000 in parentheses, are 3.043 (50.7%), 1.355 (22.6%), 0.714 (11.9%), 0.482 (8.0%), 0.256 (4.3%), and 0.150 (2.5%). At first sight, missingness has not impacted the results found before with the complete data, the first two components accounting for about 73% of the variation in the original data, neither the eigenvectors have been substantially changed. The first two components for the complete data (Equations 6.2 and 6.3) have practically the same coefficients as those based on the incomplete data, as shown here:

$$Z_1 = -0.509(\text{CSTS}) - 0.508(\text{CASF}) - 0.424(\text{CLNS}) - 0.138(\text{CVEG})$$
$$- 0.376(\text{CFRT}) - 0.379(\text{COFT})$$

$$Z_2 = 0.046(\text{CSTS}) + 0.017(\text{CASF}) + 0.122(\text{CLNS}) - 0.711(\text{CVEG})$$
$$- 0.460(\text{CFRT}) + 0.516(\text{COFT})$$

The coefficients in Z_2 might be reversed to make comparable to those for Z_2 based on the data without missing observations. Moreover, this change of sign may be necessary to produce a plot of the first two components based on the missing data (Figure 6.5), to facilitate the comparison with the plot of the same two first components shown in Figure 6.4. With a few exceptions, both plots are quite similar; the most salient variation seen in Figure 6.5 is the shift of Latvia, which now practically overlaps its position with Greece. In this example, the missing observations have the effect to concentrate the positions on the plot for countries with similar values for both components.

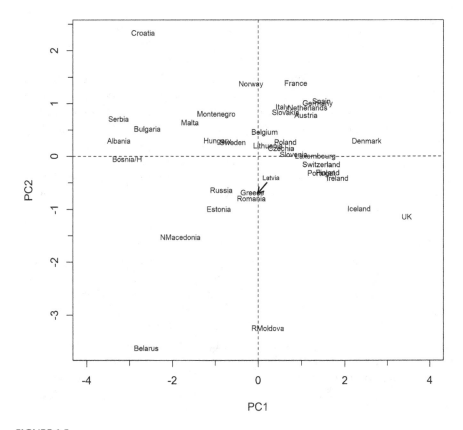

FIGURE 6.5
European countries plotted against the first two principal components for the cost of individual food groups making up a healthy diet basket, with missing observations.

6.4 Computer Programs

The appendix to this chapter provides the R code for carrying out a principal components analysis, and many standard statistical packages will carry out this analysis because it is one of the most common types of multivariate analysis in use. When the analysis is not mentioned as an option in a package, it may still be possible to do the required calculations as a special type of factor analysis, as explained in Chapter 7. In that case, care will be needed to ensure that there is no confusion between the principal components and the factors, which are the principal components scaled to have unit variances.

This confusion can also occur with some programs that claim to be carrying out a principal component analysis. Instead of providing the values of the principal components (with variances equal to eigenvalues), they provide values of the principal components scaled to have variances of one.

6.5 Further Reading

Principal components analysis is covered in almost all texts on multivariate analysis, and in greater detail by Jolliffe (2002) and Jackson (1991). The shorter monographs by Dunteman (1989) and Greenacre et al. (2022) are particularly helpful for social scientists, and applications of principal component analysis in various disciplines (e.g., biology, finance, ecology, health, and architecture) are described in Sanguansat (2012). For data scientists, the book by Kassambara (2017) provides excellent tutorials to run principal component analysis in R as well as other multivariate methods described in this book.

Exercises

1. Table 6.8 shows six measurements on each of 25 pottery goblets excavated from prehistoric sites in Thailand, with Figure 6.6 illustrating the typical shape and the nature of the measurements. The main question of interest for these data concerns similarities and differences between the goblets, with obvious questions being whether it is possible to display the data graphically to show how the goblets are related and, if so, whether there are any obvious groupings of similar goblets and any goblets that are particularly unusual. Carry out a principal components analysis and see whether the values of the principal components help to answer these questions.

TABLE 6.8

Measurements (in Centimeters) Taken on 25 Prehistoric Goblets from Thailand

Goblet	X_1	X_2	X_3	X_4	X_5	X_6
1	13	21	23	14	7	8
2	14	14	24	19	5	9
3	19	23	24	20	6	12
4	17	18	16	16	11	8
5	19	20	16	16	10	7
6	12	20	24	17	6	9
7	12	19	22	16	6	10
8	12	22	25	15	7	7
9	11	15	17	11	6	5
10	11	13	14	11	7	4
11	12	20	25	18	5	12
12	13	21	23	15	9	8

(Continued)

TABLE 6.8 (Continued)

Goblet	X_1	X_2	X_3	X_4	X_5	X_6
13	12	15	19	12	5	6
14	13	22	26	17	7	10
15	14	22	26	15	7	9
16	14	19	20	17	5	10
17	15	16	15	15	9	7
18	19	21	20	16	9	10
19	12	20	26	16	7	10
20	17	20	27	18	6	14
21	13	20	27	17	6	9
22	9	9	10	7	4	3
23	8	8	7	5	2	2
24	9	9	8	4	2	2
25	12	19	27	18	5	12

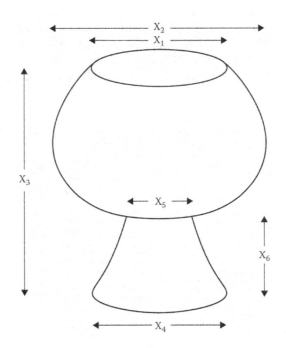

FIGURE 6.6
Measurements made on pottery goblets from Thailand.

One point that needs consideration with this exercise is the extent to which differences between goblets are due to shape differences rather than size differences. It may well be considered that two goblets that are almost the same shape but have very different sizes are really similar. The problem of separating size and shape differences has generated a considerable scientific literature, which will not be considered here. However, it can be noted that one way to remove the effects of size involves dividing the measurements for a goblet by the total height of the body of the goblet. Alternatively, the measurements on a goblet can be expressed as a proportion of the sum of all measurements on that goblet. These types of standardization of variables will ensure that the data values are similar for two goblets with the same shape but different sizes.

2. The Centers for Disease Control and Prevention, or CDC, is a service organization at the US Department of Health and Human Services (HHS) with the purpose to conduct science-based research providing health information for the protection of their citizens from health threats. The main health statistics agency in charge of data-gathering and analysis at CDC is the National Center of Health Statistics (NCHS), dependent of the US Office of Public Health Data, Surveillance, and Technology. NCHS produces data on a wide range of health indicators that have important uses for public health among the US population, like life expectancy, infant mortality, access to and use of health care services, prevalence of health conditions (such as cholesterol, hypertension, and heart diseases), leading causes of death (specific to regions, age, ethnicity, or sex groups), care quality, and patient safety. NCHS collects data from a variety of sources, in collaboration with other public and private health partners. Health indicators are classified as positive or negative, indicating whether they are associated directly or inversely with health (Pan American Health Organization, 2018). Life expectancy at birth and vaccination coverage are positive health indicators of long-term survival. On the other hand, infant or heart disease mortality rates are examples of negative indicators. Table 6.9 gives one example of the sort of data published by the NCHS, which allows to explore health indicators by state and region in the US. In this case, multivariate methods may be useful in isolating groups of states with similar patterns of health indicators, and in generally aiding the understanding of the relationships between the states and the regions. From a principal component analysis for these data, identify the determinant health indicators that may explain differences between states. Accompany the numerical results with a plot of the first two principal components, labelling points with state names.

TABLE 6.9

Health Indicators at the 50 states in the USA in the Period 2016–2018, Obtained from the National Center for Health Statistics (NCHS). The second column gives the region in which each state is allocated by the US Department of Health and Human Services (HSS).

State[1]	Region	X_1	X_2	X_3	X_4	X_5	X_6	X_7
Alabama	R4	75.1	918.1	169.8	6.94	224.7	77.3	22.8
Alaska	R10	78.0	700.3	149.4	6.25	129.7	68.8	27.0
Arizona	R9	78.7	669.2	135.2	5.71	136.4	69.9	24.9
Arkansas	R6	75.6	876.6	174.1	7.51	217.4	67.8	21.7
California	R9	80.8	609.0	137.9	4.21	139.7	65.3	27.9
Colorado	R8	80.0	651.4	131.5	4.75	124.3	76.4	29.2
Connecticut	R1	80.4	644.2	138.5	4.20	142.1	75.7	36.6
Delaware	R3	77.8	757.2	161.3	5.93	159.1	78.1	26.1
Florida	R4	78.9	657.9	144.4	6.04	143.1	67.1	27.0
Georgia	R4	77.2	790.2	156.6	7.05	175.8	77.3	23.1
Hawaii	R9	81.0	572.5	127.7	6.78	125.6	75.1	30.6
Idaho	R10	79.0	726.6	149.7	5.05	157.9	73.9	19.0
Illinois	R5	78.8	716.9	158.5	6.55	163.9	71.5	29.1
Indiana	R5	76.8	832.7	169.7	6.72	180.7	68.8	22.9
Iowa	R7	79.2	723.9	157.7	4.98	165.1	73.5	21.9
Kansas	R7	78.0	771.3	158.4	6.37	158.9	76.4	25.6
Kentucky	R4	75.3	920.0	186.6	6.05	198.3	74.5	24.4
Louisiana	R6	75.6	870.9	172.4	7.65	212.2	66.8	28.2
Maine	R1	78.6	753.9	168.0	5.52	147.0	70.6	32.4
Maryland	R3	78.5	714.1	151.5	6.02	161.9	74.4	36.9
Massachusetts	R1	80.1	670.6	146.9	4.18	131.5	85.3	44.2
Michigan	R5	77.7	782.3	162.7	6.22	195.0	70.2	30.9
Minnesota	R5	80.5	648.0	147.0	5.06	119.0	73.8	31.1
Mississippi	R4	74.6	934.8	183.7	8.41	222.1	70.4	19.8
Missouri	R7	76.6	823.2	166.3	6.35	188.4	66.9	27.8
Montana	R8	78.7	720.2	148.1	4.78	163.2	63.6	25.2
Nebraska	R7	79.1	717.4	152.7	5.77	145.7	80.6	26.0
Nevada	R9	77.9	741.1	153.7	6.14	190.7	71.9	20.6
New Hampshire	R1	79.1	712.7	153.9	3.50	151.0	78.0	31.6
New Jersey	R2	79.8	672.5	144.4	3.80	163.0	70.2	32.4
New Mexico	R6	77.2	748.7	137.9	5.69	148.2	68.5	24.2
New York	R2	80.5	626.7	140.9	4.33	171.9	72.3	37.4

(Continued)

TABLE 6.9 (Continued)

State[1]	Region	X_1	X_2	X_3	X_4	X_5	X_6	X_7
North Carolina	R4	77.6	770.1	157.8	6.75	155.5	77.8	26.0
North Dakota	R8	79.3	690.3	144.7	5.64	140.0	68.2	24.7
Ohio	R5	76.8	838.4	169.5	6.94	191.1	68.0	30.0
Oklahoma	R6	75.6	893.2	177.9	7.09	228.5	67.0	21.0
Oregon	R10	79.7	689.8	153.7	4.22	128.4	58.1	29.9
Pennsylvania	R3	78.1	759.7	160.2	5.94	176.1	73.7	33.1
Rhode Island	R1	79.8	696.7	156.1	4.95	158.9	75.5	40.7
South Carolina	R4	76.5	821.6	161.4	7.11	167.0	69.7	23.6
South Dakota	R8	78.9	715.9	154.7	5.89	156.3	70.4	24.9
Tennessee	R4	75.5	889.7	173.8	6.89	202.4	67.4	26.3
Texas	R6	78.4	731.8	146.6	5.48	170.0	69.5	23.3
Utah	R8	79.6	691.8	121.9	5.49	146.4	72.2	22.9
Vermont	R1	79.3	706.5	159.1	6.44	150.5	76.8	39.0
Virginia	R3	79.0	709.5	152.3	5.61	147.9	65.9	27.5
Washington	R10	80.0	666.6	149.5	4.69	135.4	75.7	27.3
West Virginia	R3	74.4	953.8	180.1	6.96	196.4	64.7	26.6
Wisconsin	R5	79.3	723.3	154.9	6.12	157.8	79.4	27.0
Wyoming	R8	78.1	749.6	138.6	5.33	152.7	62.8	20.2

Source: www.cdc.gov/nchs. [1] The district of Columbia has been omitted.

Note: X_1 (LEB) = Life expectancy at birth for year 2018 (Arias et al., 2021; Tejada-Vera et al., 2021), X_2 (DDA) = Death rates per 100,000 people of drug/alcohol induced causes in 2018, X_3 (DCN) = Death rate of cancer per 100,000 people in 2018, X_4 (DBI) = Linked birth/infant death rates in 2018, X_5 (DHD) = Heart disease mortality (number of deaths per 100,000 total population in 2018), X_6 (VCV) = Vaccination coverage for combined seven-vaccine series among children aged 19–35 months in 2016, X_7 (APH) = Active physicians in patient care (number by 10,000 resident population in 2018). Variables X_2– X_5 have been adjusted by age.

3. **R.** There are various projects engaged in the discovery of knowledge based on large and complex data sets (colloquially known as "big data"), using data mining and machine-learning methods. The software tool KEEL (Knowledge Extraction based on Evolutionary Learning; Alcalá-Fernández et al., 2009) is one of these projects, which also offers a data set repository (Alcalá-Fernández et al., 2011) for researchers with benchmarks tools to analyze the behavior of machine learning methods. Principal component analysis (PCA) has been recognized as a useful machine learning tool, and the application of PCA has already transcended to many disciplines like clinical research, where information is accessible from voluminous electronic

healthcare records (Zhang and Castelló, 2017). In this exercise, the reader will make use of the file **wisconsin.csv**, downloadable from the book's supplementary material repository. These clinical data were obtained from the KEEL data set repository and then edited to allow their manipulation and analysis using R, and it consists of nine quantitative attributes of tumors from patients who had undergone surgery for breast cancer. The study was conducted at the University of Wisconsin Hospitals, Madison, and the data also includes information of the class of the tumor detected for each patient, benign (B) or malignant (M). The purpose of this exercise is to undertake a PCA based on big data (699 records) with missing observations, using the pairwise-complete observations method. Describe each principal component and produce a plot of the first two components to characterize the patients with benign and malign tumors. Support the description of that plot in terms of *t*-tests of means and Levene tests of variances for the first two principal components (see Example 6.1).

References

Alcalá- Fernández, J., Fernández, A., Luengo, J., Derrac, J., García, S., Sánchez, L., and Herrera, F. (2011). KEEL data-mining software tool: data set repository, integration of algorithms and experimental analysis framework. *Journal of Multiple-Valued Logic and Soft Computing* 17 (2–3): 255–287.

Alcalá-Fernández, J., Sánchez, L., García, S., del Jesus, M.J., Ventura, S., Garrell, J.M., Otero, J., Romero, C., Bacardit, J., Rivas, V.M., Fernández, J.C., and Herrera, F. (2009). KEEL: a software tool to assess evolutionary algorithms to data mining problems. *Soft Computing* 13(3): 307–318.

Arias, E., Bastian, B., Xu, J.Q., and Tejada-Vera, B. (2021). *U.S. State Life Tables, 2018 National Vital Statistics Reports*. Hyattsville, MD: National Center for Health Statistics.

Dray, S., and Josse, J. (2015). Principal component analysis with missing values: a comparative survey of methods. *Plant Ecology* 216: 657–667.

Dunteman, G.H. (1989). *Principal Components Analysis*. Newbury Park, CA: SAGE.

FAO. (2023). *FAOSTAT. Cost and Affordability of a Healthy Diet (CoAHD)*. www.fao.org/faostat/en/#data/CAHD (Accessed: 27 March 2023).

Greenacre, M., Groenen, P.J.F., Hastie, T., Iodice d'Enza, A., Markos, A., and Tuzhilina, E. (2022). Principal component analysis. *Nature Reviews Methods Primers* 2: 100. https://doi.org/10.1038/s43586-022-00184-w.

Hotelling, H. (1933). Analysis of a complex of statistical variables into principal components. *Journal of Educational Psychology* 24: 417–441, 498–520.

Jackson, J.E. (1991). *A User's Guide to Principal Components*. New York: Wiley.

Jolliffe, I.T. (2002). *Principal Component Analysis*. 2nd Edn. New York, NY: Springer-Verlag.

Kassambara, A. (2017). *Practical Guide to Principal Component Methods in R.* Marseille: STHDA. www.sthda.com.

Pan American Health Organization. (2018). *Health Indicators. Conceptual and Operational Considerations.* Washington, DC: PAHO.

Pearson, K. (1901). On lines and planes of closest fit to a system of points in space. *Philosophical Magazine* 2: 557–572.

Podani, J., Kalapos, T., Bartra, B., and Schmera, D. (2021). Principal component analysis of incomplete data – a simple solution to an old problem. *Ecological Informatics* 61: 101235. https://doi.org/10.1016/j.ecoinf.2021.101235.

Sanguansat, P. (Ed.). (2012). *Principal Component Analysis – Multidisciplinary Applications.* Rijeka: InTech.

Tejada-Vera, B., Salant, B., Bastian, B., and Arias, E. (2021). *US Life Expectancy by State and Sex.* National Center for Health Statistics. https://www.cdc.gov/nchs/data-visualization/state-life-expectancy/index.htm

Van Ginkel, J.R. (2023). Handling missing data in principal component analysis using multiple imputation. In: van der Ark, L.A., Emons, W.H.M., and Meijer, R.R. (eds). *Essays on Contemporary Psychometrics*, pp. 141–161. Cham: Springer.

Zhang, Z., and Castelló, A. (2017). Principal components analysis in clinical studies. *Annals of Translational Medicine* 5(17): 351. https://doi.org/10.21037/atm.2017.07.12.

Appendix: Principal Components Analysis (PCA) in R

The default R installation provides two computational methods for principal components analysis, performed by two functions, `princomp` and `prcomp`, which are loaded each time R is invoked. The former uses an algorithm that closely follows the procedure described in Section 6.2, based on the calculation of eigenvalues of the correlation matrix, or the covariance matrix if this is desired. In matrix algebra terminology, this technique is known as the *spectral decomposition* of the covariance or correlation matrix. On the other hand, `prcomp` applies a method called the *singular value decomposition (SVD)* (Feeman, 2023), a general procedure of matrix factorization, useful for handling matrices that are singular or nearly singular. The application of SVD to principal component analysis is supported by two facts: first, that the non-zero singular values of any real matrix $\mathbf{M}$ are the square roots of the nonzero eigenvalues of both $\mathbf{MM}^T$ and $\mathbf{M}^T\mathbf{M}$, the product of a matrix by its transpose, and vice versa; second, that the covariance or the correlation matrices can be expressed as the multiplication of these two matrices. In general, `princomp` and `prcomp` produce similar results. However, the implementation of the SVD algorithm is numerically more accurate. Thus, `prcomp` or any other R function for principal component analysis based on the SVD procedure is preferable (Beaton et al., 2014).

R scripts are provided in this book's website as computational aids for getting the results in Example 6.1, the analysis of the Bumpus sparrow data, and Example 6.2, the analysis of health indicators in the US. The basic command used in these examples is

```
pca.results <- prcomp(data, scale = TRUE, ...)
```

The option `scale = TRUE` means that the principal components are computed on the correlation matrix. It is worth noticing that the object `pca.results` produced by `prcomp` contains the singular values of the correlation or the covariance matrix. These values, labeled with the heading *Standard Deviation*, are revealed when the function `print(pca.results)` is executed. The eigenvalues are simply these standard deviations squared. In addition to the singular values, the eigenvectors, and the values for each principal component, two-dimensional graphical summaries of the PCA can be built by means of the function `plot`, like the plots shown in Figures 6.1 and 6.2. It is also possible to produce a variation of the plot for the first two principal components through the command (offered by the default `stats` package).

```
biplot(pca.results)
```

A *biplot* (Gower and Hand, 1996; Gower et al., 2011) is a graphical summary of the PCA in which the first two principal components, plotted as points, are simultaneously displayed with a projection of the variables in the two-dimensional reduced space, plotted as arrows. Improved extensions of the `biplot` function have been implemented in packages `FactoMiner` (Le et al., 2008), `ExPosition` (Beaton et al., 2014), and `factoextra` (Kassambara and Mundt, 2020). Also, several R packages offer functions for principal component analysis in addition to `princomp` and `prcompr`. A non-exhaustive list of some of these packages and functions is given here. The one to use can be chosen based on the descriptions provided for each package in the corresponding help files or manuals. Moreover, a self-learning collection of vignettes have been implemented in package `LearnPCA` (Hanson and Harvey, 2024) for beginners wanting to understand principal component analysis better.

Additional packages have been created to perform principal component analysis with missing data. The strategy described in this book relies on the calculation of correlation coefficients based on the pairwise-complete observations method. These correlations can be computed by inserting the option `use = "pairwise.complete.obs"` to the `cor` function listed in the appendix for Chapter 2. Alternatively, the `Hmisc` package provides a fuller output including "effective" sample sizes and significance level, in addition to the correlation matrix. Once this matrix is obtained, principal component analysis (PCA) is generated by means of the `eigen` function. Podani et al. (2021) provide a simple user-function (`InDaPCA`, PCA of Incomplete Data) to perform all these steps at once; this function is exemplified in one of the scripts accompanying this chapter.

The application of PCA based on data imputation is out of the scope of this book but there are good references on the topic; three of them have been listed in Section 6.3. For R users, the `missMDA` package (Josse and Husson,

Package[1]	Function for PCA	Reference
`stats`	`princomp`	R documentation (R Core Team, 2024)
`stats`	`prcomp`	R documentation (R Core Team, 2024)
`FactoMineR`	`PCA`	Le et al. (2008)
`ade4`	`dudi.pca`	Dray and Dufour (2007)
`vegan`	`rda`	Oksanen et al. (2024)
`amap`	`acp`	Lucas (2022)
`ExPosition`	`epPCA`	Beaton et al. (2014)
`h2o`	`h2o.prcomp`	Fryda et al. (2024)
`PCAtools`	`pca`	Blighe and Lun (2024)

Source: [1] `PCAtools` is hosted at the Bioconductor project (www.bioconductor.org/); the remaining packages are located at CRAN (https://cran.r-project.org/web/packages/).

2016) is strongly recommended. This package not only provides a computational solution of PCA for missing data using multiple imputation, but it also gives a brief literature review of algorithms available to handle missing data for multivariate methods, PCA included. For more advanced R users, *R-miss-tastic* (Mayer et al., 2022), accessible from https://rmisstastic.netlify.app/, is the most complete platform offering documents, data, tutorials, and workflows for analyses with missing data based on multiple data imputation.

References

Beaton, D., Fatt, C.R.C., and Abdi, H. (2014). An ExPosition of multivariate analysis with the singular value decomposition in R. *Computational Statistics & Data Analysis* 72: 176–189.

Blighe, K., and Lun, A. (2024). *PCAtools: everything principal components analysis*. R package version 2.16.0. https://bioconductor.org/packages/PCAtools.

Dray, S., and Dufour, A.B. (2007). The ade4 package: implementing the duality diagram for ecologists. *Journal of Statistical Software* 22(4): 1–20.

Feeman, T.G. (2023). *Applied Linear Algebra and Matrix Methods*. Cham: Springer.

Fryda, T., LeDell, E., Gill, N., Aiello, S., Fu, A., Candel, A., Click, C., Kraljevic, T., Nykodym, T., Aboyoun, P., Kurka, M., Malohlava, M., Poirier, S., and Wong, W. (2024). *H2o: R Interface for the 'H2O' Scalable Machine Learning Platform*. R package version 3.44.0.3. https://CRAN.R-project.org/package=h2o.

Gower, J.C., and Hand, D.J. (1996). *Biplots*. Monographs on Statistics and Applied Probability 54. London: Chapman & Hall.

Gower, J.C., Lubbe, S., and le Roux, N. (2011). *Understanding Biplots*. Chichester: Wiley.

Hanson, B., and Harvey, D. (2024). *LearnPCA: Functions, Data Sets and Vignettes to Aid in Learning Principal Components Analysis (PCA)*. R package version 0.3.4. https://CRAN.R-project.org/package=LearnPCA.

Josse, J., and Husson, F. (2016). missMDA: a package for handling missing values in multivariate data analysis. *Journal of Statistical Software* 70(1): 1–31. https://doi.org/10.18637/jss.v070.i01.

Kassambara, A., and Mundt, F. (2020). *factoextra: Extract and Visualize the Results of Multivariate Data Analyses*. R package version 1.0.7. https://CRAN.R-project.org/package=factoextra.

Le, S., Josse, J., and Husson, F. (2008). FactoMineR: an R package for multivariate analysis. *Journal of Statistical Software* 25(1): 1–18.

Lucas, A. (2022). *amap: Another Multidimensional Analysis Package*. R package version 0.8-19. https://CRAN.R-project.org/package=amap.

Mayer, I., Sportisse, A., Tierney, N., Vialaneix, N., and Josse, J. (2022). R-miss-tastic: a unified platform for missing values methods and workflows. *The R Journal* 14(2): 244–266. https://doi.org/10.32614/RJ-2022-040.

Oksanen, J., Simpson, G., Blanchet, F., Kindt, R., Legendre, P., Minchin, P., O'Hara, R., Solymos, P., Stevens, M., Szoecs, E., Wagner, H., Barbour, M., Bedward, M., Bolker, B., Borcard, D., Carvalho, G., Chirico, M., De Caceres, M., Durand, S.,

Evangelista, H., FitzJohn, R., Friendly, M., Furneaux, B., Hannigan, G., Hill, M., Lahti, L., McGlinn, D., Ouellette, M., Ribeiro Cunha, E., Smith, T., Stier, A., Ter Braak, C., and Weedon, J. (2024). *vegan: Community Ecology Package*. R package version 2.6-6.1. https://CRAN.R-project.org/package=vegan.

Podani, J., Kalapos, T., Bartra, B., and Schmera, D. (2021). Principal component analysis of incomplete data – a simple solution to an old problem. *Ecological Informatics* 61: 101235. https://doi.org/10.1016/j.ecoinf.2021.101235.

R Core Team. (2024). *R: A Language and Environment for Statistical Computing*. Vienna: R Foundation for Statistical Computing. www.r-project.org/.

7

Factor Analysis

7.1 The Factor Analysis Model

Factor analysis has similar aims to principal components analysis. The basic idea is to describe a set of p variables $X_1, X_2, ...X_p$, in terms of a smaller number of indices or factors, and in the process get a better understanding of the relationship between these variables. There is one important difference. Principal components analysis is not based on any particular statistical model, whereas factor analysis is based on a model.

Charles Spearman is credited with the early development of factor analysis. He studied the correlations between students' test scores of various types and noted that many observed correlations could be accounted for by a simple model (Spearman, 1904). For example, in one case, he obtained the matrix of correlations shown in Table 7.1 for how boys in a preparatory school scored on tests in classics, French, English, mathematics, discrimination of pitch, and music. He noted that this matrix has the interesting property that any two rows are almost proportional if the diagonals are ignored. Thus, for rows classics and English, there are the ratios

$$\frac{0.83}{0.67} \approx \frac{0.70}{0.64} \approx \frac{0.66}{0.54} \approx \frac{0.63}{0.51} \approx 1.2.$$

Based on this observation, Spearman suggested that the six test scores could be described by the equation $X_i = a_i F + e_i$,

where

X_i is the ith standardized score (mean zero SD one) for all the boys,

a_i is a constant,

F is a factor value, with mean zero and standard deviation of one for all the boys, and

e_i is the part of X_i that is specific to the ith test only.

Spearman showed that a constant ratio between the rows of a correlation matrix follows as a consequence of these assumptions, and therefore, this is a plausible model for the data.

   DOI: 10.1201/9781003453482-7

Apart from the constant correlation ratios, it also follows that the variance of X_i is given by

$$
\begin{aligned}
Var(X_i) &= Var(a_i F + e_i) \\
&= Var(a_i F) + Var(e_i) \\
&= a_i^2 Var(F) + Var(e_i) \\
&= a_i^2 + Var(e_i)
\end{aligned}
$$

because a_i is a constant, F and e_i are assumed to be independent, and the variance of F is assumed to be unity. Also, because $Var(X_i) = 1$,

$$
a_i^2 + Var(e_i) = 1.
$$

Hence, the constant a_i, which is called the *factor loading*, is such that its square is the proportion of the variance of X_i that is accounted for by the factor.

Based on f his work, Spearman formulated his two-factor theory of mental tests. According to this theory, each test result is made up of two parts: one that is common to all the tests (general intelligence) and another that is specific to the test in question. Later, this theory was modified to allow each test result to consist of a part due to several common factors plus a part specific to the test. This gives the general factor analysis model, which states that

$$
X_i = a_{i1}F_1 + a_{i2}F_2 + \dots + a_{im}F_m + e_i = \sum_{j=1}^{m} a_{ij}F_j + e_i; \quad i = 1, 2, \dots, p,
$$

where

X_i is the ith test score with mean zero and unit variance, a_{i1} to a_{im} are the factor loadings for the ith test;

F_1 to F_m are m uncorrelated common factors, each with mean zero and unit variance; and

e_i is specific only to the ith test, is uncorrelated with any of the common factors, and has zero mean.

With this model,

$$
\begin{aligned}
Var(X_i) &= 1 = a_{i1}^2 Var(F_1) + a_{i2}^2 Var(F_2) + \dots + a_{im}^2 Var(F_m) + Var(e_i) \\
&= a_{i1}^2 + a_{i2}^2 + \dots + a_{im}^2 + Var(e_i) = \sum_{j=1}^{m} a_{ij}^2 + Var(e_i),
\end{aligned}
$$

TABLE 7.1

Correlations between Test Scores for Boys in a Preparatory School

	Classics	French	English	Mathematics	Discrimination of pitch	Music
Classics	1.00	0.83	0.78	0.70	0.66	0.63
French	0.83	1.00	0.67	0.67	0.65	0.57
English	0.78	0.67	1.00	0.64	0.54	0.51
Mathematics	0.70	0.67	0.64	1.00	0.45	0.51
Discrimination of pitch	0.66	0.65	0.54	0.45	1.00	0.40
Music	0.63	0.57	0.51	0.51	0.40	1.00

Source: Data from Spearman, C., *Am. J. Psychol.*, 15, 201–93, 1904.

where

$\sum_{j=1}^{m} a_{ij}^2$ is called the *communality* of X_i (the part of its variance related to the common factors) and

$Var(e_i)$ is called the *specificity* of X_i (the part of its variance that is unrelated to the common factors).

It can also be shown that the correlation between X_i and X_j is

$$r_{ij} = a_{i1}a_{j1} + a_{i2}a_{j2} + \ldots + a_{im}a_{jm} = \sum_{k=1}^{m} a_{ik}a_{jk}.$$

Hence, two test scores can only be highly correlated if they have high loadings on the same factors. Furthermore, $-1 \le a_{ij} \le 1$, as the communality cannot exceed one.

7.2 Procedure for a Factor Analysis

The data for a factor analysis have the same form as for a principal components analysis. That is, there are p variables with values for these n individuals, as shown in Table 6.2.

We are about to launch into a dizzying array of subscripted entities, which could be confusing to those with only a passing familiarity with summation notation. These include (definitions to follow):

- $X_i; i = 1, 2, ..., p$, the p variables measured on each of n subjects; here, the variables have been standardized:

$$X_i = \frac{Y_i - \bar{Y}}{s_Y}$$

where Y_i is the variable on its original scale, while $\bar{Y}$ and s_Y are the sample mean and standard deviation in the n measurements.

- $F_j; j = 1, 2, ..., m, m \leq p$, the so-called *provisional factors*. The number of them, m, is chosen by the researcher; typically, m is quite less than p.
- a_{ij}, the *factor loadings* for factor j on variable i.
- e_i, a random error term associated with X_i.
- $F_k^*; k = 1, 2, ..., m$, formed from linear combinations of the F_j, with weights labeled as d_{kj}.
- $Z_l; l = 1, 2, ..., p$, the p principal components[1] of the variables X_i. In practice some smaller number (m here) are retained for use.

We will now unpack all these entities to examine certain details of factor analysis.

There are three stages to factor analysis. To start, provisional factor loadings a_{ij} are determined. One approach starts with a principal components analysis and neglects the principal components after the first m, which are then taken to be the m factors. The factors found in this way are then uncorrelated with each other and are also uncorrelated with the specific factors. However, the specific factors are not uncorrelated with each other, which means that one of the assumptions of the factor analysis model does not hold. This may not matter much, providing that the communalities are high.

However they are formed, the provisional factor loadings are not unique. If $F_j; j = 1, 2, ..., m$, are the provisional factors, then linear combinations of these of the form

$$F_k^* = \sum_{j=1}^{m} d_{kj} F_j; k = 1, 2, ..., m$$

can be constructed that are uncorrelated and explain the data just as well as the provisional factors. Indeed, there are an infinite number of alternative solutions for the factor analysis model. This leads to the second stage in the analysis, which is called *factor rotation*. At this stage, the provisional factors are transformed to find new factors that are easier to interpret. To rotate or to transform in this context means essentially to choose the d_{kj} values in the presented equations.

The last stage of an analysis involves calculating the factor scores. These are the values of the rotated factors F_k^* ; $k = 1, 2, ..., m$ for each of the n individuals for which data are available.

The number of factors (m) is up to the user, although it may sometimes be suggested by the nature of the data. When a principal components analysis is used to find a provisional solution, a rough rule of thumb involves choosing m to be the number of eigenvalues greater than unity for the correlation matrix of the test scores. The logic here is the same as was explained in the previous chapter on principal components analysis. A factor associated with an eigenvalue of less than unity accounts for less variation in the data than one of the original test scores. In general, increasing m will increase the communalities of variables. However, communalities are not changed by factor rotation.

Factor rotation can be orthogonal or oblique. With orthogonal rotation, the new factors are uncorrelated, like the provisional factors. With oblique rotation, the new factors are correlated. Whichever type of rotation is used, it is desirable that the factor loadings for the new factors should be either close to zero or very different from zero. A near-zero a_{ij} means that X_i is not strongly related to the factor F_j. A large positive or negative value of a_{ij} means that X_i is determined by F_j to a large extent. If each test score is strongly related to some factors but not related to the others, then this makes the factors easier to identify than would otherwise be the case.

One method of orthogonal factor rotation that is often used is called *varimax rotation*. This assumes that the interpretability of factor j can be measured by the variance of the squares of its factor loadings, that is, the variance of $\sum a_{ij}^2$ If this variance is large, then the a_{ij} values tend to be either close to zero or close to unity. The varimax rotation, therefore, maximizes the sum of these variances for all of the factors. Kaiser (1958) first suggested this approach. Later, he modified it slightly by normalizing the factor loadings before maximizing the variances of their squares, because this appears to give improved results. Varimax rotation can, therefore, be carried out with or without Kaiser normalization. Numerous other methods for orthogonal rotation have been proposed. However, varimax rotation seems to be a good standard approach.

Sometimes, factor analysts are prepared to give up the idea of the factors being uncorrelated so that the factor loadings should be as simple as possible. An oblique rotation may then give a better solution than an orthogonal one. Again, there are numerous methods available to do the oblique rotation.

A method for calculating the factor scores for individuals based on principal components is described in the following section. There are other methods available, so the one to be used will depend on the computer package or R code being used for an analysis.

7.3 Principal Components Factor Analysis

It has been remarked that one way to do a factor analysis is to begin with a principal components analysis and use the first few principal components as unrotated factors. This has the virtue of simplicity, although as the specific factors $e_1, e_2, ..., e_p$ are correlated, the factor analysis model is not quite correct. Sometimes, factor analysts do a principal components factor analysis first and then try other approaches afterward.

The method for finding the unrotated factors is as follows. With p variables, there will be the same number of principal components. These are linear combinations of the original variables

$$Z_l = \sum_{i=1}^{p} b_{li} X_i; l = 1, 2, ..., p \tag{7.1}$$

where the b_{li} values are given by the eigenvectors of the correlation matrix. This transformation from X values to Z values is orthogonal so that the inverse relationship is simply

$$X_i = \sum_{l=1}^{p} b_{li} Z_l; i = 1, 2, ..., p.$$

For a factor analysis, only m of the principal components are retained, so the last equations become

$$X_i = \sum_{j=1}^{m} b_{ij} Z_j + e_i; i = 1, 2, ..., p; m \le p$$

where e_i is a linear combination of the principal components Z_{m+1} to Z_p. All that needs to be done now is to scale the principal components Z_l to have unit variances, as required for factors. To do this, each must be divided by its standard deviation, which is $\sqrt{\lambda_j}$ where λ_j is the corresponding eigenvalue in the correlation matrix. The equations then become

$$X_i = \sum_{j=1}^{m} \sqrt{\lambda_j} b_{ji} F_j + e_i; i = 1, 2, ..., p; m \le p.$$

Letting $F_i = Z_i / \sqrt{\lambda_i}$ and $a_{ij} = \sqrt{\lambda_i} b_{ji}$, the unrotated factor model can be written as

$$X_i = \sum_{j=1}^{m} a_{ij} F_i + e_i \,; i = 1, 2, ..., p; \, m \le p. \tag{7.2}$$

After a varimax or other type of rotation, a new solution has the form

$$X_i = \sum_{j=1}^{m} g_{ji} F_j^* + e_i \,; i = 1, 2, ..., p; \, m \le p. \tag{7.3}$$

where F_j^* represents the new *j*th factor.

The values of the *i*th unrotated factor are just the values of the *i*th principal component after these have been scaled to have a variance of one. The values of the rotated factors are more complicated to obtain, but it can be shown that these are given by the matrix equation

$$F^* = XG(G'G)^{-1}$$

F* is an $n \times m$ matrix containing the values for the m rotated factors in its columns, with one row for each of the n original rows of data.

X is the $n \times p$ matrix of the original data for p variables and n observations, after coding the variables X_1 to X_p to have means of zero and variances of one.

G is the $p \times m$ matrix of rotated factor loadings given by Equation 7.3.

7.4 Using a Factor Analysis Program to Do Principal Components Analysis

Because many computer programs for factor analysis allow the option of using principal components as initial factors, it is possible to use the programs to do principal components analysis. Extract the same number of factors as variables and do not do any rotation. The factor loadings will then be as given by Equation 7.2, with $m = p$ and $e_i = 0; i = 1, ..., p$. The principal components are given by Equation 7.1.

7.4.1 Example 7.1: Cost of Healthy Food in European Countries

In Example 6.2, a principal components analysis was carried out on cost of a healthy diet in European countries in 2017 (Table 1.5). It is of some interest to continue the examination of these data using a factor analysis model.

The correlation matrix for the six variables is given in Table 6.5, and the eigenvalues and eigenvectors of this correlation matrix are shown in Table 7.2. There are two eigenvalues greater than unity, thus suggesting that two factors should be considered, which is what will be done here.

The eigenvectors in Table 7.2 give the coefficients of the X variables for Equation 7.1.

These eigenvectors are changed into factor loadings for two factors using Equation 7.2, to give the model:

$$X_1 = +0.88 \cdot F_1 - 0.05 \cdot F_2 + e_1 (0.77)$$

$$X_2 = +0.88 \cdot F_1 - 0.02 \cdot F_2 + e_1 (0.77)$$

$$X_3 = +0.75 \cdot F_1 - 0.16 \cdot F_2 + e_1 (0.59)$$

$$X_4 = +0.22 \cdot F_1 + 0.82 \cdot F_2 + e_1 (0.73)$$

$$X_5 = +0.63 \cdot F_1 + 0.55 \cdot F_2 + e_1 (0.70)$$

$$X_6 = +0.69 \cdot F_1 - 0.56 \cdot F_2 + e_1 (0.79)$$

Here, the values in parentheses are the communalities. For example, the communality for variable X_1 is $(0.88)^2 + (-0.55)^2 \approx 0.77$. The communalities are high for all variables, with X_3 (cost of legumes, nuts, and seeds) having the smallest value of 0.59 among all variables, which is acceptable. Most of the variance for the six variables is accounted for by the two common factors.

TABLE 7.2

Eigenvalues and Eigenvectors for the Data of Cost of Healthy Food in Europe

		Eigenvectors					
PC	Eigenvalue	X_1 CSTS	X_2 CASF	X_3 CLNS	X_4 CVEG	X_5 CFRT	X_6 COFT
1	**3.023**	−0.505	−0.505	−0.432	−0.127	−0.362	−0.395
2	**1.322**	−0.041	0.017	−0.139	0.716	0.479	−0.487
3	0.729	−0.345	0.002	0.435	−0.533	0.532	−0.354
4	0.486	0.055	−0.541	0.730	0.313	−0.267	−0.034
5	0.258	−0.576	−0.208	−0.048	0.208	0.330	0.685
6	0.183	−0.538	0.639	0.264	0.215	−0.419	−0.103

Note: The variables are costs (purchasing power parity dollar per person per day) of individual food groups that make up the healthy diet basket: CSTS, starchy staples; CASF, animal source foods; CLNS, legumes, nuts, and seeds; CVEG; vegetables; CFRT, fruits; COFT, oils and fats.

Factor loadings that are 0.50 or more (ignoring the sign) are bold in these equations. These large and moderate loadings indicate how the variables are related to the factors. It can be seen that X_1 and X_2 are almost entirely accounted for by factor 1 alone, and X_3 still is largely accounted for by the same factor 1. In contrast, X_4 is accounted for by factor 2. Finally, X_5 and X_6 are a mixture of factors 1 and 2. An undesirable property of this choice of factors is that these two latter X variables are related strongly to the two of the factors. This suggests that a factor rotation may provide a simpler model for the data.

A varimax rotation with Kaiser normalization was carried out. This produced the model:

$$X_1 = +\mathbf{0.84} \cdot F_1 + 0.26 \cdot F_2 + e_1$$

$$X_2 = +\mathbf{0.82} \cdot F_1 + 0.32 \cdot F_2 + e_1$$

$$X_3 = +\mathbf{0.76} \cdot F_1 + 0.11 \cdot F_2 + e_1$$

$$X_4 = -0.08 \cdot F_1 + \mathbf{0.85} \cdot F_2 + e_1$$

$$X_5 = +0.40 \cdot F_1 + \mathbf{0.73} \cdot F_2 + e_1$$

$$X_6 = +\mathbf{0.84} \cdot F_1 - 0.29 \cdot F_2 + e_1$$

The communalities are unchanged, and the factors are still uncorrelated. However, this is a slightly better solution than the previous one, as none of the variables is dependent on more than one factor.

At this stage, it is usual to try to put labels on factors. In the present case, this is not too difficult, based on the highest loadings only.

Factor 1 has a high positive loading for X_1 (starchy staples), X_2 (animal source food), X_3 (legumes, nuts and seeds) and X_6 (oils and fats). Therefore, it measures the extent to which people pay for "non-produce" food, assuming that produce food is described as those fresh farm-produced crops for human feeding (basically fruits and vegetables), generally in the same state as where and when they were harvested. Therefore, as factor 2 has a high positive loading for X_4 (vegetables) and X_5 (fruits), its natural label is "produce food."

The **G** matrix of Equations 7.3 and 7.4 is given by the factor loadings shown in the second model in this example. For example, $g_{11} = 0.84$ and $g_{12} = 0.26$, to two decimal places. Using these loadings and carrying out the matrix calculations shown in Equation 7.4 provides the values for the factor scores for each of the 30 countries in the original data set. These factor scores are shown in Table 7.3.

From the factor scores, it can be seen that the values for factor 1 emphasize the high cost (most likely for reduced accessibility) of non-produce food in Belarus, Albania, Bosnia, and Bulgaria. In contrast, the cluster of non-produce food-types is less expensive for people living in United Kingdom and

TABLE 7.3

Rotated Factor Scores for 38 European Countries

Country	Factor 1	Factor 2
Albania	1.682	0.943
Austria	−0.797	0.375
Belarus	2.585	−2.407
Belgium	−0.502	0.235
Bosnia and Herzegovina	1.690	0.632
Bulgaria	1.265	0.974
Croatia	0.763	2.471
Czechia	−0.288	−0.002
Denmark	−1.398	−0.326
Estonia	0.876	−0.679
Finland	−0.713	−0.622
France	−0.873	0.951
Germany	−1.006	0.502
Greece	0.431	−0.545
Hungary	0.459	0.435
Iceland	−0.897	−1.346
Ireland	−0.817	−0.758
Italy	−0.780	0.645
Latvia	0.098	0.111
Lithuania	−0.146	0.116
Luxembourg	−0.832	−0.369
Malta	0.690	0.852
Montenegro	0.311	0.860
Netherlands	−1.092	0.399
North Macedonia	1.477	−0.819
Norway	−0.309	1.150
Poland	−0.380	0.078
Portugal	−0.646	−0.586
Republic of Moldova	0.934	−2.715
Romania	0.374	−0.629
Russian Federation	0.666	−0.268
Serbia	1.577	1.256
Slovakia	−0.562	0.543
Slovenia	−0.411	−0.158
Spain	−1.082	0.527

(Continued)

TABLE 7.3   (Continued)

Country	Factor 1	Factor 2
Sweden	0.083	0.266
Switzerland	−0.970	−0.385
United Kingdom	−1.460	−1.706

Note: Factor 1 is "non-produce food," and factor 2 is "produce food."

Denmark. The values for factor 2 indicate that Croatia stands out among the European countries with higher costs of produce food, followed by Serbia and Norway. On the other hand, people from Moldova and Belarus pay less for fruits and vegetables in comparison to other European countries. All these assertions can be verified by revisiting the original database, paying attention to the groups of variables determined by each factor, and comparing values of the original variables.

For other data sets involving several variables it would be possible and reasonable to continue the factor analysis by trying models with fewer factors and different methods of factor extraction. However, sufficient things have been said already to indicate the general approach, so the example will be left at this point.

It should be kept in mind by anyone attempting to reproduce this analysis that different statistical packages may give the eigenvectors shown in Table 7.3 except that all the coefficients have their signs reversed. A sign reversal may also occur through a factor rotation so that the loadings for a rotated factor are the opposite of what is shown.

7.5 Options in Analyses

Computer programs for factor analysis, including different R codes, may allow a number of different options for the analysis, which is likely to be rather confusing for the novice in this area. Typically, there might be four or five methods for the initial extraction of factors and about the same number of methods for rotating these factors (including no rotation). This gives in the order of 20 different types of factor analysis that can be carried out, with results that will differ to some extent at least.

There is also the question of the number of factors to extract. Some packages may make an automatic choice, which may or may not be acceptable. The possibility of trying different numbers of factors, therefore, increases the choices for an analysis even more. Among practitioners who apply factor analysis in psychological, social, and educational sciences, there has been a consensus in the strategy called *parallel analysis* (PA; Horn, 1965) as a tool to determine nontrivial factors from observed data. PA represents a variation of the Kaiser's rule described in this chapter, by adding a randomization step based on the idea that interpretable key factors should have eigenvalues greater than parallel components derived from random data having the same sample size and number of variables (Hayton et al., 2004). Factors corresponding to eigenvalues larger than the parallel average random eigenvalues are retained. Extensions to this basic algorithm include comparing the observed eigenvalues with the 95th percentile of the distribution of randomly generated factor eigenvalues to decide whether the factor is retained or not.

On the whole, it is probably best to avoid using too many options when first practicing factor analysis. The use of principal components as initial factors with varimax rotation, as used in the example in this chapter, is a reasonable start with any set of data. The maximum likelihood method for extracting factors is a good approach in principle and might also be tried if this is available.

7.6 The Value of Factor Analysis

Factor analysis is something of an art, and it is certainly not as objective as many statistical methods. For this reason, some statisticians are skeptical about its value. For example, Chatfield and Collins (1986) list six problems with factor analysis and conclude that "factor analysis should not be used in most practical situations." Similarly, Seber (2004) notes as a result of simulation studies that even if the postulated factor model is correct, then the chance of recovering it using available methods is not high.

On the other hand, factor analysis is widely used to analyze data and, no doubt, will continue to be widely used in the future. The reason for this is that users find the results useful for gaining insight into the structure of multivariate data. Therefore, if it is thought of as a purely descriptive tool, with limitations that are understood, it must take its place as one of the important multivariate methods. What should be avoided is carrying out a factor analysis on a single small sample that cannot be replicated and then assuming that the factors obtained must represent underlying variables that exist in the real world.

7.7 Discussion and Further Reading

Although, as noted in the previous section, the topic of factor analysis covered in this chapter is sometimes not presented enthusiastically (Chatfield and Collins, 1986; Seber, 2004), there are several books exclusively devoted to the topic (Garson, 2023; Watkins, 2021; Holmes Finch, 2020; Gorsuch, 2015;

Fabrigar and Wegener, 2012), and the method is presented in many texts on multivariate analysis (e.g., Afifi et al., 2020; Everitt and Hothorn, 2011; Rencher, 2002). All these books are generally positive about the applicability of factor analysis. For example, Rencher (2002) discusses at length the validity of factor analysis and why it often fails to work. He notes that there are many sets of data for which factor analysis should not be used, but others for which the method is useful.

Factor analysis as discussed in this chapter is often referred to as *exploratory factor analysis* because it starts with no assumptions about the number of factors that exist or the nature of these factors. In this respect, it differs from what is called *confirmatory factor analysis*, which requires the number of factors and the factor structure to be specified in advance. In this way, confirmatory factor analysis can be used to test theories about the structure of the data. Confirmatory factor analysis is more complicated to carry out than exploratory factor analysis. The details are described by Roos and Bauldry (2022), Everitt and Hothorn (2011), and Tabachnick and Fidell (2019). Confirmatory factor analysis is a special case of structural equation modeling, which is covered in chapter 14 of Tabachnick and Fidell (2019); see also Loehlin (2004).

Exercise

Consider the US NCHS data about health indicators, introduced in Chapter 6, Exercise 2. Carry out a factor analysis of the data, giving a name to each factor chosen, and examine the relationship between states with respect to these factors.

Note

1. In keeping with usual subscript labelling practice of selecting letters in sequence from the alphabet (usually starting with i, as we do here), this subscript is a lowercase L.

References

Afifi, A., May, S., Donatello, R.A., and Clark, V.A. (2020). *Practical Multivariate Statistics*. 6th Edn. Boca Raton, FL: CRC Press.

Chatfield, C., and Collins, A.J. (1986). *Introduction to Multivariate Analysis*. London: Chapman and Hall.

Everitt, B., and Hothorn, T. (2011). *An Introduction to Applied Multivariate Analysis with R.* New York: Springer.

Fabrigar, L.R., and Wegener, D.T. (2012). *Exploratory Factor Analysis.* Oxford: Oxford University Press.

Garson, G.D. (2023). *Factor Analysis and Dimension Reduction in R.* London: Routledge.

Gorsuch, R.L. (2015). *Factor Analysis. Classic Edition.* New York: Routledge.

Hayton, J.C., Allen, D.G., and Scarpello, V. (2004). Factor retention decisions in exploratory factor analysis: a tutorial on parallel analysis. *Organizational Research Methods* 7(2): 191–205.

Holmes Finch, W. (2020). *Exploratory Factor Analysis.* Los Angeles, CA: SAGE.

Horn, J.L. (1965). A rationale and a test for the number of factors in factor analysis. *Psychometrika* 30: 179–185.

Kaiser, H.F. (1958). The varimax criterion for analytic rotation in factor analysis. *Psychometrika* 23: 187–200.

Loehlin, J.C. (2004). *Latent Variable Models. An Introduction to Factor, Path, and Structural Equation Analysis.* Mahwah, NJ: Lawrence Erlbaum Associates, Inc., Publishers.

Rencher, A.C. (2002). *Methods of Multivariate Analysis.* Wiley Series in Probability and Statistics. New York: Wiley.

Roos, J.M., and Bauldry, S. (2022). *Confirmatory Factor Analysis.* Los Angeles, CA: SAGE.

Seber, G.A.F. (2004). *Multivariate Observations.* New York: Wiley.

Spearman, C. (1904). "General intelligence", objectively determined and measured. *American Journal of Psychology* 15: 201–293.

Tabachnick, B.G., and Fidell, L.S. (2019). *Using Multivariate Statistics.* 7th Edn. New York: Pearson.

Watkins, M.W. (2021). *A Step by Step Guide to Exploratory Factor Analysis with R and R Studio.* New York: Routledge.

Appendix: Factor Analysis in R

In its default package `stats`, R offers the function `factanal` as a maximum likelihood (ML) method for extracting factors, a topic noted briefly in Section 7.5. Thus, ML factor analysis is also considered the default factor analysis in R. However, in Section 7.5, it was also emphasized that there are different approaches in factor analysis, each approach associated with a particular algorithm. Psychometric researchers have been the most interested in applying the range of existing algorithms for factor analysis. This explains why the R package `psych`, created and maintained by Revelle (2024), has been considered as the main tool for psychometric applications of several multivariate methods, including factor analysis. The `psych` package offers the `fa` function, from which the user may choose one of seven algorithms of factor analysis (minimum residual, principal axes, alpha factoring, weighted least squares, generalized least squares, minimum rank, and maximum likelihood factor analysis). Nevertheless, none of these options follows exactly the algorithm described in Section 7.3, whereby a principal components analysis (PCA) is used to produce initial factors, followed by a varimax rotation and the calculation of factor scores, which are also known as Bartlett scores, using Equation 7.4.

It is not difficult to execute most of the steps of PCA factor analysis with the set of R functions already considered in previous chapters (e.g., `prcomp`, matrix multiplication, and matrix inversion). The particular step completing this algorithm, varimax rotation with Kaiser normalization, can be performed with the `varimax` function implemented in the `stats` package. However, this way of doing a PCA factor analysis can be avoided with `principal`, another function in the `psych` package. Although this function is thought just to be doing a PCA, its output is organized in such a way that the component loadings are more suitable for a typical factor analysis, showing the best m factors. The developer of `psych` argues that the presence of `principal` in his package as a choice for factor analysis, in addition to the algorithms executed by the `fa` function, is because "psychologists typically use PCA in a manner similar to factor analysis and thus the principal function produces output that is perhaps more understandable than that produced by `princomp` in the `stats` package" (Revelle, 2017). The command required to replicate the factor analysis described in Chapter 7 is then

```
principal(data, nfactors = 2, rotate = "varimax")
```

At this book's website, the reader will find two R scripts written to carry out the factor analysis performed for Example 7.1. One script does the calculations in the fastest way via the `principal` function. The second script makes use of functions `prcomp` and `varimax`. It has been written for instructive purposes so that the reader can follow in detail the application of Equations 7.1 through 7.4 in this chapter.

For more advanced R users wanting to apply the algorithms implemented in `pysch` as indicated, in combination with additional procedures and R-packages for factor analysis (e.g., parallel analysis), the books by Garson (2023) and Watkins (2021) are worth reading.

References

Garson, G.D. (2023). *Factor Analysis and Dimension Reduction in R*. London: Routledge.

Revelle, W. (2017). *An Overview of the Psych Package: Vignette of Psych Procedures for Psychological, Psychometric, and Personality Research*. http://personality-project.org/r/overview.pdf.

Revelle, W. (2024). *psych: Procedures for Psychological, Psychometric, and Personality Research*. Evanston, IL: Northwestern University. R package version 2.4.3. https://CRAN.R-project.org/package=psych.

Watkins, M.W. (2021). *A Step by Step Guide to Exploratory Factor Analysis with R and R Studio*. New York: Routledge.

8

Discriminant Function Analysis

8.1 The Problem of Separating Groups

The problem that is addressed with discriminant function analysis is the extent to which it is possible to separate two or more groups of individuals, given measurements for these individuals on several variables. For example, with the data in Table 1.1 on five body measurements of 21 surviving and 28 nonsurviving sparrows, it is interesting to consider whether it is possible to use the body measurements to separate survivors and nonsurvivors. Also, for the data shown in Table 1.2 on four dimensions of Egyptian skulls for samples from five time periods, it is reasonable to consider whether the measurements can be used to assign skulls to different time periods.

In the general case, there will be m random samples from different groups, with sizes $n_1, n_2, \ldots, n_m$, and values will be available for p variables $X_1, X_2, \ldots, X_p$ for each sample member. Thus, the data for a discriminant function analysis takes the form shown in Table 8.1; each datum x_{ijk} represent the observed ith value in group j for the kth variable, where $i = 1, 2, \ldots, n_j$,

TABLE 8.1

The Form of Data for a Discriminant Function Analysis with m Groups with Possibly Different Sizes $n_1, n_2, \ldots, n_m$, and p Variables Measured on Each Individual Case

Case	X_1	X_2	$\cdots$	X_p	Group
1	x_{111}	x_{112}	$\cdots$	x_{11p}	1
2	x_{211}	x_{212}	$\cdots$	x_{21p}	1
$\vdots$	$\vdots$	$\vdots$	$\vdots$	$\vdots$	$\vdots$
n_1	x_{n_111}	x_{n_112}	$\cdots$	x_{n_11p}	1
1	x_{121}	x_{122}	$\cdots$	x_{12p}	2
2	x_{221}	x_{222}	$\cdots$	x_{22p}	2

(Continued)

DOI: 10.1201/9781003453482-8

TABLE 8.1   (Continued)

Case	X_1	X_2	$\cdots$	X_p	Group
$\vdots$	$\vdots$	$\vdots$	$\vdots$	$\vdots$	
n_2	$x_{n_2 21}$	$x_{n_2 22}$	$\cdots$	$x_{n_2 2p}$	2
$\vdots$	$\vdots$	$\vdots$	$\vdots$	$\vdots$	$\vdots$
1	x_{1m1}	x_{1m2}	$\cdots$	x_{1mp}	m
2	x_{2m1}	x_{2m2}	$\cdots$	x_{2mp}	m
$\vdots$	$\vdots$	$\vdots$	$\vdots$	$\vdots$	$\vdots$
n_m	$x_{n_m m1}$	$x_{n_m m2}$	$\cdots$	$x_{n_m mp}$	m

$$\sum_{j=1}^{m} n_j = N$$

$j = 1, 2, \ldots, m$, $k = 1, 2, \ldots, p$. In principle, the data for a discriminant function analysis do not need to be standardized to have zero means and unit variances prior to the start of the analysis, as is usual with principal components and factor analysis. This is because the outcome of a discriminant function analysis is not affected in any important way as far as the scaling of individual variables do not differ too much.

8.2 Discrimination Using Mahalanobis Distances

One approach to discrimination is based on Mahalanobis distances, as defined in Section 5.3. The mean vectors for the m samples can be regarded as estimates of the true mean vectors for the groups. The Mahalanobis distances from the individual cases to the group centers can then be calculated, and each individual can be allocated to the group to which it is closest. This may or may not be the group that the individual actually came from, so the percentage of correct allocations is an indication of how well groups can be separated using the available variables.

This procedure is more precisely defined as follows. Let

$$\bar{x}_j' = (\bar{x}_{j1}, \bar{x}_{j2}, \ldots, \bar{x}_{jp})$$

denote the vector of mean values for the sample from the jth group, let C_j denote the covariance matrix for the same sample, and let C denote the

pooled sample covariance matrix, where these vectors and matrices are calculated as explained in Section 2.7. Then, the Mahalanobis distance from an
observation $\mathbf{x}' = (x_1, x_2, \ldots, x_p)$ to the center of group j is estimated to be

$$D_j^2 = (\mathbf{x} - \bar{\mathbf{x}}_j)'\mathbf{C}^{-1}(\mathbf{x} - \bar{\mathbf{x}}_j)$$

$$\sum_{r=1}^{p}\sum_{s=1}^{p}(x_r - \bar{x}_{rj})c_{(rs)}(x_s - \bar{x}_{sj}) \tag{8.1}$$

where $c_{(rs)}$ is the element in the rth row and the sth column of $\mathbf{C}^{-1}$. The observation $\mathbf{x}$ is then allocated to the group for which D_j^2 has the smallest value.

8.3 Canonical Discriminant Functions

It is sometimes useful to be able to determine functions of the variables
$X_1, X_2, \ldots, X_p$ that in some sense separate the m groups as much as is possible.
The simplest approach then involves taking a linear combination of the X
variables

$$Z = a_1 X_1 + a_2 X_2 + \cdots + a_p X_p$$

for this purpose. Groups can be well separated using values of Z if the mean
value of this variable changes considerably from group to group, with the
values within a group being fairly constant.

One way to determine the coefficients $a_1, a_2, \ldots, a_p$ in the index involves
choosing these so as to maximize the F-ratio for a one-way analysis of variance. Thus, if there are a total of N individuals in all the groups, an analysis of variance on Z values takes the form shown in Table 8.2 (compare to
Table 4.4). Hence, a suitable function for separating the groups can be defined
as the linear combination for which the F-ratio M_B / M_W is as large as possible, as first suggested by Fisher (1936).

TABLE 8.2

An Analysis of Variance on the Z index

Source of variation	Df	Mean square	F-ratio
Between groups	$m-1$	M_B	M_B / M_W
Within groups	$N-m$	M_W	
Total	$N-1$		

When this approach is used, it turns out that it may be possible to determine several linear combinations for separating groups. In general, the number available, e.g., s, is the smaller of p and $m-1$ or, symbolically, $s = \min(p, m-1)$. The linear combinations are referred to as *canonical discriminant functions*.

The first function,

$$Z_1 = a_{11}X_1 + a_{12}X_2 + \cdots + a_{1p}X_p,$$

gives the maximum possible F-ratio for a one-way analysis of variance for the variation within and between groups. If there is more than one function, then the second one,

$$Z_2 = a_{21}X_1 + a_{22}X_2 + \cdots + a_{2p}X_p,$$

gives the maximum possible F-ratio on a one-way analysis of variance, subject to the condition that there is no correlation between Z_1 and Z_2 within groups. Further functions are defined in the same way. Thus, the lth canonical discriminant function

$$Z_l = a_{l1}X_1 + a_{l2}X_2 + \cdots + a_{lp}X_p; \, l = 1, 2, ..., s = \min(p, m-1)$$

is the linear combination for which the F-ratio on an analysis of variance is maximized, subject to Z_l being uncorrelated with $Z_1, Z_2, \ldots,$ and Z_{l-1} within groups.

Finding the coefficients of the canonical discriminant functions turns out to be an eigenvalue problem. The within-sample matrix of sums of squares and cross products, $\mathbf{W}$, and the total sample matrix of sums of squares and cross products matrix, $\mathbf{T}$, are calculated as described in Section 4.7. From these, the between-groups matrix

$$\mathbf{B} = \mathbf{T} - \mathbf{W}$$

can be determined. Next, the eigenvalues and eigenvectors of the matrix $\mathbf{W}^{-1}\mathbf{B}$ have to be found. If the eigenvalues are $\lambda_1 > \lambda_2 > ... > \lambda_s$, then λ_l is the ratio of the between-group sum of squares to the within-group sum of squares for the lth linear combination, Z_l, while the elements of the corresponding eigenvector $\mathbf{a}_l' = (a_{l1}, a_{l2}, ..., a_{lp})$ are the coefficients of the X variables for this index. The calculation of eigenvectors is not as immediate as in principal components and factor analysis because constraints are imposed to the solution. A transformation of the data is applied to guarantee that any eigenvector $\mathbf{a}_l$ is scaled in such a way that it satisfies the condition $\mathbf{a}_l'\mathbf{C}\mathbf{a}_l = 1$, where $\mathbf{C} = \mathbf{W}/(N-m)$ is the estimated within-groups pooled covariance matrix **C**. To do so, the observed $N \times p$ data matrix **X** is first transformed into $\mathbf{Y} = \mathbf{T}^{-1}\mathbf{X}$, where $\mathbf{W} = \mathbf{TT}^{-1}$ is called the Cholesky decomposition of **W** (here, **T** is lower triangular square matrix, i.e., all elements above the diagonal are zero). Then the within-groups sum of squares for the Y-variables will be the identity matrix, thus imposing the so-called *sphericity condition* for the transformed within-groups covariance matrix which, in turn, implies the fulfillment of the constraint $\mathbf{a}_l'\mathbf{Ca}_l = 1$. The between-groups sum of squares matrix for **Y** will be $\mathbf{B}_Y = \mathbf{T}^{-1}\mathbf{B}(\mathbf{T}^{-1})'$, where **B** is between-groups matrix defined earlier. It can be proved that the eigenvalues of $\mathbf{B}_Y$ are the same as those for $\mathbf{W}^{-1}\mathbf{B}$, and the eigenvectors $\mathbf{a}_l$ of $\mathbf{W}^{-1}\mathbf{B}$ are obtained from the eigenvectors $\mathbf{b}_{(Y)l}$ of $\mathbf{B}_Y$ using the equation

$$\mathbf{a}_l = (\mathbf{T}')^{-1}\mathbf{b}_{(Y)l}\sqrt{N-m}, l = 1,2,\ldots,s \qquad (8.2).$$

The elements of each scaled eigenvector (8.2) making up the canonical discriminant function Z_l are called *raw canonical coefficients* for Z_l. In practice, the raw coefficients and the original X variables should be used to compute the discriminant scores (Klecka, 1980). However, when making significant interpretations with canonical discriminant functions and the X variables are not measured on the same scale or their variances are not comparable, it is recommended to use standardized instead of raw coefficients (Rencher, 2002). Only in this situation the X variables might be explicitly standardized before discriminant analysis. Nonetheless, this data-standardization step can be omitted as the new *standardized canonical coefficients* are obtained just from the raw coefficients and the diagonal elements of the pooled sample covariance matrix, **C**. Symbolically, if $\mathbf{a}_l' = (a_{l1},a_{l2},\ldots,a_{lp})$ is the set of raw coefficients corresponding to the lth discriminant function Z_l and $c_{11},c_{22},\ldots,c_{pp}$ are the diagonal elements of **C**, then the lth set of standardized coefficients, $\mathbf{d}_l$, are

$$\mathbf{d}_l' = (d_{l1},d_{l2},\ldots,d_{lp}) = \left(\sqrt{c_{11}}a_{l1},\sqrt{c_{22}}a_{l2},\ldots,\sqrt{c_{pp}}a_{lp}\right).$$

Take into account that the interpretation of the canonical discriminant functions based on the standardized coefficients must be done using the standardized X variables.

$Z_1,Z_2,\ldots,Z_s$ are linear combinations of the original variables such that Z_1 reflects group differences as much as possible, Z_2 captures as much as possible of the group differences not displayed by Z_1, Z_3 captures as much as possible of the group differences not displayed by Z_1 and Z_2, and so on. The hope is that the first few functions are sufficient to account for almost all the important group differences. In particular, if only the first one or two functions are needed for this purpose, then a simple graphical representation of the relationship between the various groups is possible by plotting the values of these functions for the sample individuals. Although the Z_l values to be plotted can be computed using the original X variables, it is customary to center these latter around the overall multivariate sample mean before mapping the discriminant functions onto the plot. Thus, if $\mathbf{X}_{cent} = \mathbf{X} - \bar{\mathbf{X}}$ is the data matrix of centered variables around the overall mean $\bar{\mathbf{X}}$ and $\mathbf{a}_l$ is the *l*th eigenvector (column vector) containing raw canonical coefficients for Z_l, then the canonical function values (or the *l*th *canonical scores*) for the observed sample are given by the elements of the *l*th column of

$$\mathbf{X}_{cent}\,\mathbf{a}_l.$$

The two-dimensional plot may also include (group) *centroids*, showing the position of the mean canonical scores corresponding to each group, as a visual aid in the interpretation of the ability of the functions on the plot to discriminate between groups.

8.4 Tests of Significance

Several tests of significance are useful in conjunction with a discriminant function analysis. In particular, the T^2-test of Section 4.3 can be used to test for a significant difference between the mean values for any pair of groups, while one of the tests described in Section 4.7 can be used to test for overall significant differences between the means for the m groups.

In addition, a test is sometimes used for testing whether the mean of the discriminant function Z_l differs significantly from group to group. This is based on the individual eigenvalues of the matrix $\mathbf{W}^{-1}\mathbf{B}$. For example, sometimes, the statistic

$$\varphi_l^2 = [N-1-(p+m)/2]\log_e(1+\lambda_l)$$

is used, where N is the total number of observations in all groups. This statistic is then tested against the chi-squared distribution with $p+m-2l$ degrees of freedom (df), and a significantly large value is considered to provide evidence that the population mean values of Z_l vary from group to group. Alternatively, the sum $\varphi_l^2 + \varphi_{l+1}^2 + \cdots + \varphi_s^2$ is sometimes used for testing for group differences related to discriminant functions Z_l to Z_s. This is tested against the chi-squared distribution, with the df being the sum of those associated with the component terms. Other tests of a similar nature are also used.

Unfortunately, these tests are suspect to some extent because the lth discriminant function in the population may not appear as the lth discriminant function in the sample due to sampling errors. For example, the estimated first discriminant function (corresponding to the largest eigenvalue for the sample matrix $\mathbf{W}^{-1}\mathbf{B}$) may in reality correspond to the second discriminant function for the population being sampled. Simulations indicate that this can upset the chi-squared tests described in the previous paragraph quite seriously. Therefore, it seems that the tests should not be relied on to decide how many of the obtained discriminant functions represent real group differences. See Harris (2001) for an extended discussion of the difficulties surrounding these tests and alternative ways to examine the nature of group differences.

One useful type of test that is valid, at least for large samples, involves calculating the Mahalanobis distance from each of the observations to the mean vector for the group containing the observation, as discussed in Section 5.3. These distances should follow approximately chi-squared distributions with p df. Hence, if an observation is very significantly far from the center of its group in comparison with the chi-squared distribution, then this brings into question whether the observation really came from the group in question.

8.5 Assumptions

The methods discussed so far in this chapter are based on two assumptions. First, for all the methods, the population within-group covariance matrix should be the same for all groups. Second, for tests of significance, the data should be multivariate normally distributed within groups.

In general, it seems that multivariate analyses that assume normality may be upset quite badly if this assumption is not correct. This contrasts with the situation with univariate analyses such as regression and analysis of variance, which are generally quite robust to this assumption. However, a failure of one or both assumptions does not necessarily mean that a discriminant function analysis is a waste of time. For example, it may well turn out that excellent discrimination is possible on data from nonnormal distributions, although it may not then be simple to establish the statistical significance of the group differences. Furthermore, discrimination methods that do not require the equality of population covariance matrices are available, as discussed in Section 8.12.

8.5.1 Example 8.1: Comparison of Samples of Egyptian Skulls

This example concerns the comparison of the values for $p = 4$ measurements on male Egyptian skulls for $m = 5$ samples ranging in age from the early predynastic period (circa 4000 BC) to the Roman period (circa AD 150). The data are shown in Table 1.2, and it has already been established that the mean values differ significantly from sample to sample (Example 4.4), with the differences tending to increase with the time difference between samples (Example 5.3). Mahalanobis distances from individual cases to the group centers can be explored first. According to the notation introduced in Section 8.2, the first observation

Case	X_1	X_2	X_3	X_4	Period
1	$x_{111} = 131$	$x_{112} = 138$	$x_{113} = 89$	$x_{114} = 49$	Early predynastic

defines a vector $\mathbf{x}' = (131, 138, 89, 49)$ and Table 5.3 can be used to create the mean vectors for each period:

$$\bar{\mathbf{x}}_1' = (131.4, 133.6, 99.2, 50.5),$$
$$\bar{\mathbf{x}}_2' = (132.4, 132.7, 99.1, 50.2),$$
$$\bar{\mathbf{x}}_3' = (134.5, 133.8, 96.0, 50.6)$$

$$\bar{\mathbf{x}}_4' = (135.5, 132.3, 94.5, 52.0), \quad \bar{\mathbf{x}}_5' = (136.2, 130.3, 93.5, 51.4),$$

All these vectors together with the covariance matrix shown in Table 5.3 provide the essential pieces to compute the Mahalanobis distance function of Equation 8.1 D_j^2 between the first observation and the jth period, $j = 1, 2, 3, 4, 5$:

$$D_1^2 = 6.55, \ D_2^2 = 7.00, \ D_3^2 = 4.33, \ D_4^2 = 5.33, \ D_5^2 = 6.17.$$

If the 150 skulls are allocated to the samples to which they are closest according to D_j^2, then only 51 of them (34%) are allocated to the samples to which they really belong (Table 8.3). It seems that skulls dimensions are not too effective in aging skulls using the discrimination method based on Mahalanobis distances.

The canonical discrimination functions applied to these data is tried now to explore an additional approach to separate periods. The within-sample and total sample matrices of sums of squares and cross products are calculated as described in Section 4.7. They are found to be

$$\mathbf{W} = \begin{bmatrix} 3061.07 & 5.33 & 11.47 & 291.30 \\ 5.33 & 3405.27 & 754.00 & 412.53 \\ 11.47 & 754.00 & 3505.97 & 164.33 \\ 291.30 & 412.53 & 164.33 & 1472.13 \end{bmatrix}$$

TABLE 8.3

Results Obtained When 150 Egyptian Skulls Are Allocated to the Group for Which They Have the Minimum Mahalanobis Distance

Source period	Early predynastic	Late predynastic	12th/13th Dynasty	Ptolemaic period	Roman period	Total
Early predynastic	12	8	4	4	2	30
Late predynastic	10	8	5	4	3	30
12th/13th Dynasty	4	4	15	2	5	30
Ptolemaic period	3	3	7	5	12	30
Roman period	2	4	4	9	11	30

and

$$\mathbf{T} = \begin{bmatrix} 3563.89 & -222.81 & -615.16 & 426.73 \\ -222.81 & 3635.17 & 1046.28 & 346.47 \\ -615.16 & 1046.28 & 4309.26 & -16.40 \\ 426.73 & 346.47 & -16.40 & 1533.33 \end{bmatrix}.$$

The between-sample matrix is, therefore,

$$\mathbf{B} = \mathbf{T} - \mathbf{W} = \begin{bmatrix} 502.83 & -228.15 & -626.63 & 135.43 \\ -228.15 & 229.91 & 292.28 & -66.07 \\ -626.63 & 292.28 & 803.29 & -180.73 \\ 135.43 & -66.07 & -180.73 & 61.20 \end{bmatrix}.$$

The eigenvalues of $\mathbf{W}^{-1}\mathbf{B}$ are found to be $\lambda_1 = 0.425$, $\lambda_2 = 0.039$, $\lambda_3 = 0.016$, and $\lambda_4 = 0.002$, and the corresponding canonical discriminant functions expressed in terms of raw canonical coefficients, are

$$Z_1 = -0.1267X_1 + 0.0370X_2 + 0.1451X_3 - 0.0829X_4$$

$$Z_2 = 0.0387X_1 + 0.2101X_2 - 0.0681X_3 - 0.0773X_4$$

$$Z_3 = 0.0928X_1 - 0.0246X_2 + 0.0147X_3 - 0.2964X_4 \tag{8.3}$$

and

$$Z_4 = -0.1488X_1 + 0.0004X_2 - 0.1325X_3 - 0.0669X_4.$$

Because λ_1 is much larger than the other eigenvalues, it is apparent that most of the sample differences are described by Z_1 alone.

The X variables in Equation 8.3 refers to the whole set of four equations for Z_1, Z_2, Z_3, and Z_4 are the values as shown in Table 1.2 without standardization. The nature of the variables is illustrated in Figure 1.1, from which it can be seen that large values of Z_1 correspond to skulls that are tall but narrow, with long jaws and short nasal heights.

The Z_1 values for individual skulls are calculated in the obvious way, using the discriminant function with raw coefficients. For example, the first skull in the early predynastic sample has $X_1 = 131$mm, $X_2 = 138$mm, $X_3 = 89$, and $X_4 = 131$mm. Therefore, for this skull

$$Z_1 = -0.1267 \times 131 + 0.0370 \times 138 + 0.1451 \times 89 - 0.0829 \times 49 = -2.640$$

The means and standard deviations found for the Z_1 values for the five samples are shown in Table 8.4. The mean of Z_1 has become lower over time, indicating a trend toward shorter, broader skulls with short jaws but relatively large nasal heights. This is, however, very much an average change. Thus, although this discriminant function analysis has been successful in pinpointing the changes in skull dimensions over time, it has not produced a satisfactory method for aging individual skulls, as it was concluded based on the Mahalanobis distances (Table 8.3).

8.5.2 Example 8.2: Discriminating between Groups of European Countries

The data shown in Table 1.5 on the cost of six food groups for 38 European countries have already been examined by principal components analysis and by factor analysis (Examples 6.2 and 7.1). Here, they will be considered from the point of view of the extent to which it is possible to discriminate between groups of countries on the basis of patterns of food-type costs. In particular, four groups existed in the period when the data were collected. These were (1) the countries that joined the European Union until 1995 (EU95), namely, Austria, Belgium, Denmark, Finland, France, Germany, Greece, Ireland, Italy, Luxembourg, the Netherlands, Portugal, Spain, Sweden, and the United Kingdom; (2) the countries that joined the EU between 1996 and 2017, specifically Bulgaria, Croatia, Czechia, Estonia, Hungary, Latvia, Lithuania, Malta, Poland, Romania, Slovakia, and Slovenia; (3) members of the European Economic Area, the non-EU countries of Iceland and Norway; and (4) the remaining non-EU countries of Albania, Belarus, Bosnia, Montenegro, North

TABLE 8.4

Means and Standard Deviations for the Discriminant Function Z_1 with Five Samples of Egyptian Skulls

Sample (period)	Mean	Standard deviation
Early predynastic	−1.489	1.167
Late predynastic	−1.638	0.842
12th and 13th Dynasties	−2.332	0.889
Ptolemaic period	−2.852	0.937
Roman period	−3.109	1.122

Macedonia, Moldova, Russia, Serbia, and Switzerland. These four groups can be used as a basis for a discriminant function analysis. Wilks' lambda test (Section 4.7) gives a very highly significant result (p-value < 0.001), so there is very clear evidence, overall, that these groups are meaningful. The number of canonical variables is three in this example, this being the minimum of the number of variables ($p = 6$) and the number of groups minus one ($m − 1 = 3$). The eigenvalues of $\mathbf{W}^{-1}\mathbf{B}$ corresponding to the three canonical variables are $\lambda_1 = 2.122$, $\lambda_2 = 0.476$, and $\lambda_3 = 0.203$. The first canonical variable is, therefore, clearly the most important. The canonical variables expressed in terms of raw coefficients are

$$Z_1 = +0.1101\,CSTS + 7.4812\,CASF + 0.9414\,CLNS + \\ 0.3202\,CVEG - 3.7182\,CFRT + 27.8932\,COFT$$

$$Z_2 = +11.6432\,CSTS - 0.5027\,CASF - 2.9398\,CLNS + \\ 0.3179\,CVEG - 7.0395\,CFRT - 26.7389\,COFT$$

$$Z_3 = -1.8798\,CSTS + 1.0304\,CASF + 9.0120\,CLNS + \\ 4.5438\,CVEG - 3.2266\,CFRT - 16.3245\,COFT$$

Different computer programs are likely to output these canonical variables with all the signs reversed for the coefficients of one or more of the variables. Also, it is important to note that it is the original food-type costs that are used in these equations rather than these costs after they have been standardized to have zero means and unit variances. However, the evident difference in the order of magnitude among raw coefficients, with variable *COFT* dominating in all of canonical discriminant functions, indicates that standardized coefficients should be preferred for interpretation. The corresponding discriminant functions with standardized coefficients are as follows:

$$Z_1 = +0.0110\,CSTS + 0.8918\,CASF + 0.0991\,CLNS +$$
$$0.0572\,CVEG - 0.4834\,CFRT + 0.5689\,COFT$$

$$Z_2 = +1.1442\,CSTS - 0.0599\,CASF - 0.3094\,CLNS +$$
$$0.0568\,CVEG - 0.9153\,CFRT - 0.5453\,COFT$$

$$Z_3 = -0.1847\,CSTS + 0.1228\,CASF + 0.9486\,CLNS +$$
$$0.8119\,CVEG - 0.4195\,CFRT - 0.3329\,COFT$$

As an additional aid for interpretation of these canonical functions, it is helpful to this respect to consider the correlations between the original variables and the canonical variables, as shown in Table 8.5. If standardized variables are chosen instead of the original ones, the correlations are identical.

From the relative magnitude of the standardized coefficients, it is seen that the first canonical variable discriminate groups of countries with high and low costs of animal food, oils, and fats; the large correlations of $CASF$ (cost of animal source food) and $COFT$ (cost of oils and fat) with Z_1 support this interpretation. Although the correlations between Z_1 and, both, $CSTS$ (cost of starchy staples) and $CLNS$ (the cost of legumes, nuts, and seeds) are above 0.5, these variables do not have the capability to discriminate among groups of countries. $CFRT$ (cost of fruits) might have a discriminant role among groups according to its standardized coefficient in Z_1, but its low correlation with the first canonical function justifies its exclusion from the interpretation.

There are not large positive or negative correlations between the second canonical variable and the original variables. The only discriminant variable seems to be $CFRT$, the cost of fruits, explained by the fact that its standardized canonical coefficient is large and its correlation with the second

TABLE 8.5

Correlations between the Standardized Food Costs and the Three Canonical Variates

Variable	Z_1	Z_2	Z_3
CSTS	0.75	0.32	0.21
CASF	0.85	-0.13	0.13
CLNS	0.52	-0.29	0.57
CVEG	-0.03	0.16	0.65
CFRT	0.18	-0.52	0.28
COFT	0.88	-0.11	-0.16

canonical variable is above 0.5. Finally, the interpretation of the third canonical variable is much clearer, with variables *CLNS* (cost of legumes, nuts, and seeds) and *CVEG* (cost of vegetables) having an important discriminant role, on the basis of their standardized coefficients and a moderate correlation with Z_3.

Plots of the countries against their values for canonical variables 1 and 2 are shown in Figures 8.1a. and 8.1b., and for canonical variables 1 and 3, in Figures 8.2a. and 8.2b. The plot of the second canonical variable against the first (Figure 8.1a) shows a distinction between most non-EU countries (NEU) at the right-hand side and most of the EU95 countries at the left-hand side. The EU17 countries lie at the center of the plot. The centroids in Figure 8.1a. show the average positions of groups in the plot, and they also suggest that group separation is not clear in some cases. Figures 8.1a. and b. indicate that an EU95 country, Greece, is far from the EU95 centroid, and it is more alike to non-EU countries located in Eastern Europe and the Balkans. In contrast, two non-EU countries, Montenegro and Switzerland, occupy a position more alike to EU countries. The case of the two-country group, EEA (Iceland and Norway), warrants a distinct interpretation. The projection of both their positions and their centroid with respect to the first canonical variable shows that their scores lie between the centroids for EU95 and EU17, and the projection of the EEA centroid along the second canonical variable (on the vertical axis) is the largest. EEA is a group of size two which is the smallest size allowed in discriminant analysis. Apparently, the isolated position of Norway and its large score along the second canonical variable is a consequence of the combination of two factors: a small sample size and the presence of outliers. A boxplot of its *CVEG* values can be used to confirm that the corresponding value for Norway is an outlier. Although *CVEG* is not a discriminant variable for the second canonical variable, the outlying *CVEG* value for Norway impacted its score on that canonical variable.

The plot of the third canonical variable against the first one (Figure 8.2a. and b.) shows no real vertical separation of the country groups, with the centroid projections being very close to each other. The position of Croatia, an EU17 country, is explained by the fact that *CVEG* is one of the variables governing the scores of the third canonical variable. The largest *CVEG* value among all 38 countries is attained by Croatia, thus producing the largest score for the third canonical variable.

The discriminant function analysis has been successful in this example in separating the non-EU countries from most of the EU95, EU17 and EEA countries, where a few exceptions have been also identified. However, there was less success in separating the two-country group EEA from the EU95 and EU17. The separation is perhaps clearer than what was obtained using principal components, as shown in Figure 6.2.

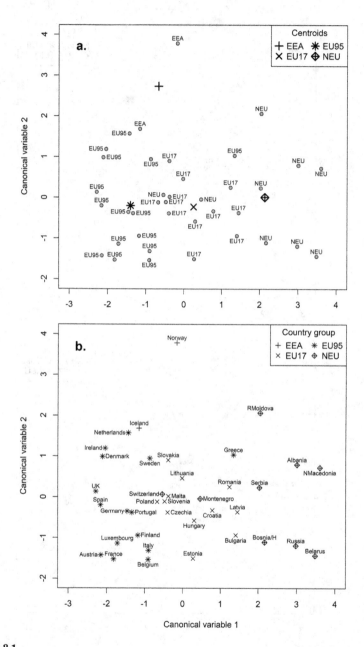

FIGURE 8.1
(a) Plot of 38 European countries against their values for the first two canonical discriminant functions, according to the group of countries given in Table 1.5. Centroid positions are highlighted for each group. (b) A similar plot as in (a), but centroids are removed, and points are labeled by country.

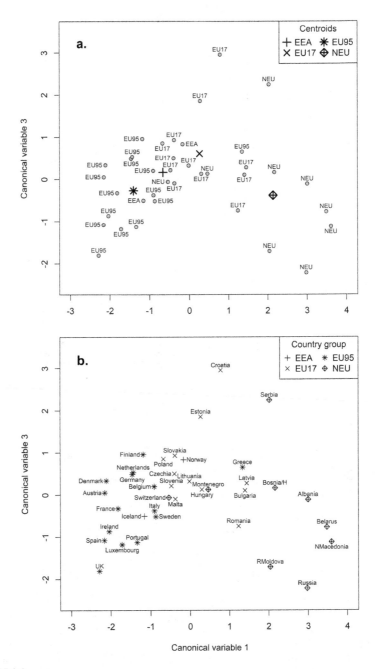

FIGURE 8.2
(a) Plot of 38 European countries against their values for the first and third canonical discriminant functions, according to the group of countries given in Table 1.5. Centroid positions are highlighted for each group. (b) A similar plot as in (a), but centroids are removed, and points are labeled by country.

8.6 Allowing for Prior Probabilities of Group Membership

Computer programs often allow many options for a discriminant function analysis. One situation is that the probability of membership is inherently different for different groups. For example, if there are two groups, it might be known that most individuals fall into group 1, while very few fall into group 2. In that case, if an individual is to be allocated to a group, it makes sense to bias the allocation procedure in favor of group 1. Thus, the process of allocating an individual to the group from which it has the smallest Mahalanobis distance should be modified. To allow for this, some computer programs permit prior probabilities of group membership to be taken into account in the analysis.

8.7 Stepwise Discriminant Function Analysis

Another possible modification of the basic analysis involves carrying it out in a stepwise manner. In this case, variables are added to the discriminant functions one by one until it is found that adding extra variables does not give significantly better discrimination. There are many different criteria that can be used for deciding which variables to include in the analysis and which to miss out.

A problem with stepwise discriminant function analysis is the bias that the procedure introduces into significance tests. Given enough variables, it is almost certain that some combination of them will produce significant discriminant functions by chance alone. If a stepwise analysis is carried out, then it is advisable to check its validity by rerunning it several times with a random allocation of individuals to groups to see how significant are the results obtained. For example, with the Egyptian skull data, the 150 skulls could be allocated completely at random to five groups of 30, the allocation being made a number of times, and a discriminant function analysis run on each random set of data. Some idea could then be gained of the probability of getting significant results through chance alone.

This type of randomization analysis to verify a discriminant function analysis is unnecessary for a standard nonstepwise analysis provided there is no reason to suspect the assumptions behind the analysis. It could, however, be informative in cases where the data are clearly not normally distributed within groups, or where the within-group covariance matrix is not the same for each group. For example, Manly and Navarro Alberto (2021, Example 12.5) shows a situation in which the results of a standard discriminant function analysis are clearly suspect by comparison with the results of a randomization analysis.

8.8 Jackknife Classification of Individuals

Using an allocation matrix such as that shown in Table 8.3 must tend to have a bias in favor of allocating individuals to the group that they really come from. After all, the group means are determined from the observations in that group. It is, therefore, not surprising to find that an observation is closest to the center of the group where the observation helped to determine that center.

To overcome this bias, some computer programs carry out what is called a *jackknife classification* of observations. This involves allocating each individual to its closest group without using that individual to help determine a group center. In this way, any bias in the allocation is avoided. In practice, there is often not a great deal of difference between the straightforward classification and the jackknife classification, with the jackknife classification usually giving a slightly smaller number of correct allocations.

8.9 Assigning Ungrouped Individuals to Groups

Some computer programs allow the input of data values for a number of individuals for which the true group is not known. It is then possible to assign these individuals to the group that they are closest to, in the Mahalanobis distance sense, on the assumption that they have to come from one of the m groups that are sampled. Obviously, in these cases, it will not be known whether the assignment is correct. However, the error in the allocation of individuals from known groups is an indication of how accurate the assignment process is likely to be. For example, the results shown in Table 8.3 indicate that allocating Egyptian skulls to different time periods using skull dimensions is liable to result in many errors.

8.10 Logistic Regression

A rather different approach to discrimination between two groups involves making use of logistic regression. To explain how this is done, the more usual use of logistic regression will first be briefly reviewed.

The general framework for logistic regression is that there are m groups to be compared, with group i consisting of n_i items, of which Y_i exhibit a positive response (a success), and $n_i - Y_i$ exhibit a negative response (a failure). The assumption is then made that the probability of a success for an item in group i is given by

$$\pi_i = \frac{\exp(\beta_0 + \beta_1 x_{i1} + \beta_2 x_{i2} + \cdots + \beta_p x_{ip})}{1 + \exp(\beta_0 + \beta_1 x_{i1} + \beta_2 x_{i2} + \cdots + \beta_p x_{ip})} \qquad (8.4)$$

where x_{ij} is the value of some variable X_j that is the same for all items in the group. In this way, the variables X_1 to X_p are allowed to influence the probability of a success, which is assumed to be the same for all items in the group, irrespective of the successes or failures of the other items in that or any other group. Similarly, the probability of a failure is $1 - \pi_i$ for all items in the ith group. It is permissible for some or all of the groups to contain just one item. Indeed, some computer programs only allow for this to be the case. There need be no concern about arbitrarily choosing what to call a success and what to call a failure. It is easy to show that reversing these designations in the data simply results in all the β values and their estimates changing sign, and consequently changing π_i into $1 - \pi_i$.

The function that is used to relate the probability of a success to the X variables is called a *logistic function*. Unlike the standard multiple regression function, the logistic function forces estimated probabilities to lie within the range zero to one. It is for this reason that logistic regression is more sensible than linear regression as a means of modeling probabilities. There are numerous computer programs available for fitting Equation 8.4 to data, that is, for estimating the values of β_0 to β_p, including R codes, as discussed in the appendix to this chapter.

In the context of discrimination with two samples, three different types of situations have to be considered:

1. The data consist of a single random sample taken from a population of items that is itself divided into two parts. The application of logistic regression is then straightforward, and the fitted Equation 8.4 can be used to give an estimate of the probability of an item being in one part of the population (i.e., being a success) as a function of the values that the item possesses for variables X_1 to X_p. In addition, the distribution of success probabilities for the sampled items is an estimate of the distribution of these probabilities for the full population.

2. Separate sampling is used, whereby a random sample of size n_1 is taken from the population of items of one type (the successes), and an independent random sample of size n_2 is taken from the population of items of the second type (the failures). Logistic regression can still be used. However, the estimated probability of a success obtained from the estimated function must be interpreted in terms of the sampling scheme and the sample sizes used.

3. Groups of items are chosen to have particular values for the variables X_1 to X_p such that these variable values change from group to group. The number of successes in each group is then observed. In this case,

the estimated logistic regression equation gives the probability of a success for an item conditional on the values that the item possesses for X_1 to X_p. The estimated function is, therefore, the same as for Situation (1), but the sample distribution of probabilities of a success is in no way an estimate of the distribution that would be found in the combined population of items that are successes or failures.

The following examples illustrate the differences between Situations (1) and (2), which are the ones that most commonly occur. Situation (3) is really just a standard logistic regression situation, and will not be considered further here.

8.10.1 Example 8.3: Storm Survival of Female Sparrows (Reconsidered)

The data in Table 1.1 consist of values for five morphological variables for 49 female sparrows taken in a moribund condition to Hermon Bumpus' laboratory at Brown University, Rhode Island, after a severe storm in 1898. The first 21 birds recovered and the remaining 28 died, and there is some interest in knowing whether it is possible to discriminate between these two groups on the basis of the five measurements. It has already been shown that there are no significant differences between the mean values of the variables for survivors and nonsurvivors (Example 4.1), although the nonsurvivors may have been more variable (Example 4.2). A principal components analysis has also confirmed the test results (Example 6.1).

This is a situation of Type (1) if the assumption is made that the sampled birds were randomly selected from the population of female sparrows in some area close to Bumpus' laboratory. Actually, the assumption of random sampling is questionable because it is not clear exactly how the birds were collected. Nevertheless, the assumption will be made for this example.

The logistic regression option in many standard computer packages can be used to fit the model

$$\pi_i = \frac{\exp(\beta_0 + \beta_1 x_{i1} + \beta_2 x_{i2} + \cdots + \beta_5 x_{i5})}{1 + \exp(\beta_0 + \beta_1 x_{i1} + \beta_2 x_{i2} + \cdots + \beta_5 x_{i5})}$$

where
  X_1 = total length (mm),
  X_2 = alar extent (mm),
  X_3 = length of beak and head (mm),
  X_4 = length of the humerus (mm),
  X_5 = length of the sternum (mm), and

π_i denotes the probability of the ith bird recovering from the storm.

TABLE 8.6

Estimates of the Constant Term and the Coefficients of X Variables When a Logistic Regression Model Is Fitted to Data on the Survival of 49 Female Sparrows

Variable	β estimate	Standard Error	Chi-squared	p-Value
Constant	13.582	15.865	—	—
Total length	−0.163	0.140	1.36	0.244
Alar extent	−0.028	0.106	0.07	0.794
Length of beak and head	−0.084	0.629	0.02	0.894
Length humerus	1.062	1.023	1.08	0.299
Length keel of sternum	0.072	0.417	0.03	0.864

Note: The chi-squared value is (estimated β value/standard error)^2. The p-value is the probability of a value this large from the chi-squared distribution with one degree of freedom. A small p-value (say, less than 0.05) provides evidence that the true value of the β parameter concerned is not equal to zero.

A chi-squared test for whether the variables account significantly for the difference between survivors and nonsurvivors gives the value 2.85 with five df, which is not at all significantly large when compared with chi-squared tables ($P = 0.722$). There is, therefore, no evidence from this analysis that the survival status was related to the morphological variables. Estimated values for β_0 to β_5 are shown in Table 8.6, together with estimated standard errors and chi-squared statistics for testing whether the individual estimates differ significantly from zero. Again, there is no evidence of any significant effects.

8.10.2 Example 8.4: Comparison of Two Samples of Egyptian Skulls

As an example of separate sampling, in which the sample size in the two groups being compared is not necessarily related in any way to the respective population sizes, consider the comparison between the first and last samples of Egyptian skulls for which data are provided in Table 1.2. The first sample consists of 30 male skulls from burials in the area of Thebes during the early predynastic period (circa 4000 BC) in Egypt, and the last sample consists of 30 male skulls from burials in the same area during the Roman period (circa AD 150). For each skull, measurements are available for X_1 = maximum breadth, X_2 = basibregmatic height, X_3 = basialveolar length, and X_4 = nasal height, all in millimeters (Figure 1.1). For the purpose of this example, it will be assumed that the two samples were effectively randomly chosen from their respective populations, although there is no way of knowing how realistic this is.

Obviously, the equal sample sizes in no way indicate that the population sizes in the two periods were equal. The sizes are, in fact, completely arbitrary because many more skulls have been measured from both periods, and an unknown number of skulls have either not survived intact or not been found. Therefore, if the two samples are lumped together and treated as a sample of size 60 for the estimation of a logistic regression equation, then it is clear that the estimated probability of a skull with certain dimensions being from the early predynastic period may not really be estimating the true probability at all.

In fact, it is difficult to define precisely what is meant by the true probability in this example, because the population is not at all clear. A working definition is that the probability of a skull with specified dimensions being from the predynastic period is equal to the proportion of all skulls with the given dimensions that are from the predynastic period in a hypothetical population of all male skulls from either the predynastic or the Roman period that might have been recovered by archaeologists in the Thebes region.

It can be shown (Seber, 2004, p. 312) that if a logistic regression is carried out on a lumped sample to estimate Equation 8.4, then the modified equation

$$\pi_i = \frac{\exp\left(\beta_0 - \log_e\left[(n_1 P_2)/(n_2 P_1)\right] + \beta_1 x_{i1} + \beta_2 x_{i2} + \cdots + \beta_p x_{ip}\right)}{1 + \exp\left(\beta_0 - \log_e\left[(n_1 P_2)/(n_2 P_1)\right] + \beta_1 x_{i1} + \beta_2 x_{i2} + \cdots + \beta_p x_{ip}\right)} \quad (8.5)$$

is what really gives the probability that an item with the specified X values is a success. Here, Equation 8.5 differs from Equation 8.4 because of the term $\log_e[(n_1 P_2)/(n_2 P_1)]$ in the numerator and the denominator, where P_1 is the proportion of items in the full population of successes and failures that are successes, and $P_2 = 1 - P_1$ is the proportion of the population that are failures. This, then, means that estimating the probability of an item with the specified X values being a success requires that P_1 and P_2 are either known or can somehow be estimated separately from the sample data to adjust the estimated logistic regression equation for the fact that the sample sizes n_1 and n_2 are not proportional to the population frequencies of successes and failures. In the example being considered, this requires that estimates of the relative frequencies of predynastic and Roman skulls in the Thebes area must be known to be able to estimate the probability that a skull is predynastic, given the values that it possesses for the variables X_1 to X_4.

Logistic regression was applied to the lumped data from the 60 predynastic and Roman skulls, with a predynastic skull being treated as a success. The resulting chi-squared test for the extent to which success is related to the X variables is 27.13 with four df. This is significantly large at the 0.1% level, giving very strong evidence of a relationship. The estimates of the constant term and the coefficients of the X variables are shown in Table 8.7. It can be seen that the estimate of β_1 is significantly different from zero at about the 1% level, and β_3 is significantly different from zero at the 2% level. Hence, X_1 and X_3 appear to be the important variables for discriminating between the two types of skull.

TABLE 8.7

Estimates of the Constant Term and the Coefficients of X variables When a Logistic Regression Model Is Fitted to Data on 30 Predynastic and 30 Roman Period Male Egyptian Skulls

Variable	β estimate	Standard Error	Chi-squared	p-Value
Constant	−6.723	13.081	—	—
Maximum breadth	−0.202	0.076	7.13	0.008
Basibregmatic height	0.129	0.079	2.66	0.103
Basialveolar length	0.177	0.073	5.84	0.016
Nasal height	−0.008	0.104	0.01	0.939

Note: See Table 8.6 for an explanation of the columns.

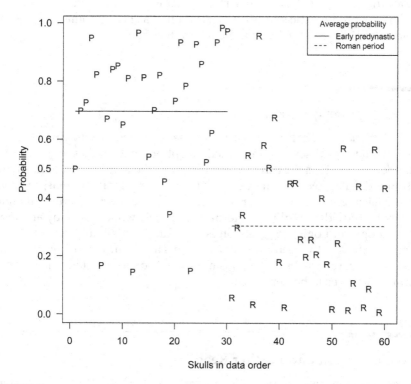

FIGURE 8.3
Values from the fitted logistic regression function plotted for 30 predynastic (P) and 30 Roman (R) skulls.

The fitted function can be used to discriminate between the two groups of skulls by assigning values for P_1 and $P_2 = 1 - P_1$ in Equation 8.5. As already noted, it is desirable that these should correspond to the population proportions of predynastic and Roman skulls. However, this is not possible, because these proportions are not known. In practice, therefore, arbitrary values must be assigned. For example, suppose that P_1 and P_1 are both set equal to 0.5. Then, $\log_e[(n_1 P_2)/(n_2 P_1)] = \log_e(1) = 0$, because $n_1 = n_2$, and Equations 8.4 and 8.5 become identical. The logistic function, therefore, estimates the probability of a skull being predynastic in a population with equal frequencies of predynastic and Roman skulls.

The extent to which the logistic equation is effective for discrimination is indicated in Figure 8.3, which shows the estimated values of π_i for the 60 sample skulls. There is a distinct difference in the distributions of values for the two samples, with the mean for predynastic skulls being about 0.7 and the mean for Roman skulls being about 0.3. However, there is also a considerable overlap between the distributions. As a result, if the sample skulls are classified as being predynastic when the logistic equation gives a value greater than 0.5 or as Roman when the equation gives a value of less than 0.5, then six predynastic skulls are misclassified as being Roman, and seven Roman skulls are misclassified as being predynastic.

8.11 Computer Programs

Major statistical packages, including R (see the appendix to this chapter), generally have a discriminant function option that applies the methods described in Sections 8.2 through 8.5, based on the assumption of normally distributed data. Because the details of the order of calculations, the way the output is given, and the terminology vary considerably, manuals may have to be studied carefully to determine precisely what is done by any individual program. Logistic regression is also fairly widely available. In some programs, there is the restriction that all items are assumed to have different values for X variables. However, it is more common for groups of items with common X values to be permitted.

8.12 Discussion and Further Reading

The assumption that samples are from multivariate distributions with the same covariance matrix, which is required for the use of the methods described in Sections 8.2 through 8.5, can be relaxed. If the samples being compared are assumed to come from multivariate normal distributions with different covariance matrices, then a method called *quadratic discriminant function analysis* can be applied. This option is also available in many computer packages. See Seber (2004, p. 297) for more information about this method and a discussion of its performance relative to the more standard linear discriminant function analysis.

Discrimination using logistic regression has been described in Section 8.10 in terms of the comparison of two groups. More detailed treatments of this method are provided by Hosmer et al. (2013) and Collett (2002). The method can also be generalized for discrimination between more than two groups if necessary, under several names, including polytomous regression. See Hosmer et al. (2013, Chapter 8) for more details. This type of analysis is a standard option in many computer packages.

Exercises

1. Consider the data in Table 4.6 for nine mandible measurements on samples from five different canine groups. Carry out a discriminant function analysis to see how well it is possible to separate the groups using the measurements.

2. Still considering the data in Table 4.6, investigate each canine group separately to see whether logistic regression shows a significant difference between males and females for the measurements. Note that in view of the small sample sizes available for each group, it is unreasonable to expect to fit a logistic function involving all nine variables with good estimates of parameters. Therefore, consideration should be given to fitting functions using only a subset of the variables.

References

Collett, D. (2002). *Modelling Binary Data*. 2nd Edn. Boca Raton, FL: Chapman and Hall/CRC.

Fisher, R.A. (1936). The utilization of multiple measurements in taxonomic problems. *Annals of Eugenics* 7: 179–188.

Harris, R.J. (2001). *A Primer on Multivariate Statistics*. 2nd Edn. New York: Psychology.

Hosmer, D.W., Lemeshow, S., and Sturdivant, R.X. (2013). *Applied Logistic Regression*. 3rd Edn. Hoboken, NJ: Wiley.

Klecka, R.W. (1980). *Discriminant Analysis*. Newbury Park, CA: SAGE.

Manly, B.F.J., and Navarro Alberto, J.A. (2021). *Randomization, Bootstrap and Monte Carlo Methods in Biology*. 4th Edn. Boca Raton, FL: Chapman and Hall/CRC.

Rencher, A.C., and Christensen, W.F. (2002). *Methods of Multivariate Analysis*. 3rd Edn. Hoboken, NJ: Wiley.

Seber, G.A.F. (2004). *Multivariate Observations*. New York: Wiley-Interscience.

Appendix: Discriminant Function Analysis in R

A.1 Canonical Discriminant Analysis in R

From a computational point of view, the method of canonical discriminant functions encompasses the eigenvalue analysis of $\mathbf{W}^{-1}\mathbf{B}$ mediated by the Cholesky decomposition of $\mathbf{W}$ described in Section 8.3, a task that can be carried out with the R-functions `eigen` (described in the Appendix for Chapter 2), `chol` (to output the triangular matrix $\mathbf{T}$ such that $\mathbf{W} = \mathbf{TT}^{-1}$), and `manova` (described in the Appendix for Chapter 4). This strategy is illustrated for Example 8.1 (the comparison of samples of Egyptian skulls) with an R script that can be downloaded from this book's website.

An alternative to using R programming is provided by the function `lda`, included in the package `MASS` (Venables and Ripley, 2002), implementing methods for linear discriminant analysis (LDA). Canonical discriminant analysis and LDA have similar goals, but LDA emphasizes more on classification rather than on testing differences among centroids. In fact, the coefficients of linear discriminants in `lda` coincide with the raw coefficients introduced in Section 8.3. Two different methods of variable specification can be given in `lda`: either

```
discan.object <- lda(Group.f ~ X1 + X2 + ... , ...)
```

or

```
discan.object <- lda(Xmat, Group.f, ...)
```

In the first method, `Group.f` is a grouping factor and the variables `X1`, `X2`, ... are the discriminator variables, reminding us that discriminant analysis involves continuous independent variables and a categorical dependent variable (i.e., the `Group.f` argument). The second option assumes that `Xmat` is a matrix or data frame whose columns are the discriminators. It is possible to specify probabilities of group membership in `lda` with the `prior` option.

The `print` method of `lda` does not produce eigenvalues. Instead, it produces singular values, which are the ratio of the between- and within-group standard deviations of the linear discriminant variables. In addition, the output of `lda` includes the *proportion of trace*, which is the proportion accounted by each eigenvalue of $\mathbf{W}^{-1}\mathbf{B}$ with respect to the sum of all the eigenvalues.

This proportion can be interpreted equivalently as the proportion of the between-group variance present on each discriminant axis.

When `lda` is executed, the user may choose to produce a classification table similar to Table 8.4 in Example 8.1 through the function predict, using

```
table(Group.f, predict(discan.object)$class)
```

The function `predict` accepts an additional parameter with the name of a data frame containing new data to be assigned to a particular group based on the canonical discriminant functions. It is worth noting that `predict` is not necessary when considering jackknifing classification of individuals, which is a procedure that can be produced in `lda` with the option CV = TRUE. Here, CV stands for cross validation, which is another name given in multivariate analysis for the jackknife classification method. An example of this command is

```
discan.object.cv <- lda(group.f ~ X1 + X2 + ... ,
                        CV = TRUE, ...)
```

The content of the object `discan.object.cv` (a list) is different from that generated by `lda` with CV = FALSE (the default). Now, `discan.object.cv` includes the vector `class`, and it is not difficult to build a table with original membership of individuals and their jackknife classification based on the discriminant analysis. See an R code exemplifying this procedure in the book's website.

The function `lda` also includes a `plot` method, which is useful for displaying one, two, or more linear discriminant functions using

```
plot(discan.object, dimen, ...)
```

The resulting plot depends on the parameter `dimen`, the number of dimensions chosen. See the R documentation for further details.

A useful alternative R function for canonical discriminant function analysis, more closely related to the methods covered in this chapter, is offered by the function `candisc`, which is present in the package with the same name (Friendly and Fox, 2024). The function `candisc` calls a multivariate linear model via the `lm` function to produce a MANOVA, the basis of the method of canonical discriminant analysis described here. `candisc` computes scaled eigenvectors that make raw and standardized canonical coefficients available, as well as correlations (also known as the *structure*) between the original variables and the canonical scores. In addition, `candisc` offers a plot method that permits the display of centroids (means) of discriminant scores for each group. R codes containing `lda` and `candisc` functions are available in the book's website, as computational aids for the discriminant analyses described in Examples 8.1 and 8.2.

A.2 Discriminant Analysis Based on Logistic Regression in R

Logistic regression is available in R as one option of the function `glm` for fitting generalized linear models (Dobson and Barnett, 2018). A typical logistic regression analysis is written as

```
logistic.model <- glm(Y ~ X1 + X2 + ...,
                      family = binomial(link = "logit"),...)
```

where

   Y is a binary response variable and

   X1, X2, ... are explanatory variables.

To use logistic regression for discriminant analysis of two samples, it is only necessary to code one sample as 1 and the other as 0 and assign these binary values to a new variable Y. The parameter estimates of the logistic regression can be obtained with the `summary` function

```
summary(logistic.model)
```

In addition, chi-squared tests for model comparison are accessible through the command `anova`. Thus,

```
anova(model.1, model.2, test = "Chisq")
```

evaluates whether the fit of `model.2` improves over the fit of `model.1`, assuming that the variables in `model.1` is a subset of the variables in `model.2`. The chi-squared tests indicated in the examples for Section 8.10 can be carried out in that way, using R: `model.1` is the intercept-only model (i.e., no variables) and `model.2` includes all the discriminators of interest. R codes are provided in the book's website exemplifying the use of `glm` and `anova` in Examples 8.3 and 8.4.

The extensions of linear discriminant analysis that are described in Sections 8.7 and 8.12 are available for the R user. The function `stepclass` located in the package `klaR` (Weihs et al., 2005) is what R offers for those interested in stepwise discriminant function analysis, while the function `qda` in the package `MASS` (Venables and Ripley, 2002) allows the separation of groups using quadratic discriminant analysis. Finally, polytomous regression, the extension of two-group logistic regression for more than two groups, is accessible through the function `multinom` from the package `nnet` (Venables and Ripley, 2002).

References

Dobson, A.J., and Barnett, A.G. (2018). *An Introduction to Generalized Linear Models*. 4th Edn. Boca Raton, FL: CRC Press.

Friendly, M., and Fox, J. (2024). *Candisc: Visualizing Generalized Canonical Discriminant and Canonical Correlation Analysis*. R package version 0.9.0. http://CRAN.R-project.org/package=heplots.

Venables, W.N., and Ripley, B.D. (2002). *Modern Applied Statistics*. 4th Edn. New York: Springer.

Weihs, C., Ligges, U., Luebke, K., and Raabe, N. (2005). klaR analyzing German business cycles. In Baier, D., Decker, R., and Schmidt-Thieme, L. (eds). *Data Analysis and Decision Support*, pp. 335–343. Berlin: Springer.

9

Cluster Analysis

9.1 Uses of Cluster Analysis

Suppose we have a sample of n objects, each of which has a numerical score on p variables. Cluster analysis can be used to devise a scheme for grouping the objects into classes so that similar objects are in the same class. The number of classes is not usually predetermined. This problem is clearly more difficult than the problem for a discriminant function analysis because with discriminant function analysis, the groups are known at the outset.

Cluster analyses are useful in many disciplines. For example, in psychiatry, with uncertainty over the classification of depressed patients, cluster analysis has been used to define objective groups. Cluster analysis may also be useful for data reduction. For example, a large number of cities can potentially be used as test markets for a new product, but it is only feasible to use a few. If cities can be placed into a small number of groups of similar cities, then one or two members from each group can be used for the test market. Alternatively, if cluster analysis generates unexpected groupings, then this might suggest relationships to be investigated.

9.2 Types of Cluster Analysis

Many algorithms have been proposed for cluster analysis. Here, attention will mostly be restricted to two particular approaches. First are hierarchical techniques that produce a dendrogram, as shown in Figure 9.1. These methods start with the calculation of the distances from each object to all other objects. Groups are then formed by a process of agglomeration or division. With agglomeration, all objects start by being alone in groups of one. Close groups are then gradually merged, till finally all objects are in a single group. With division, all objects start in a single group. This is then split into two groups, and the two groups are then split, and so on, till all objects are in groups of their own.

A second approach to cluster analysis involves partitioning, with objects being allowed to move in and out of groups at different stages of the analysis.

                    DOI: 10.1201/9781003453482-9

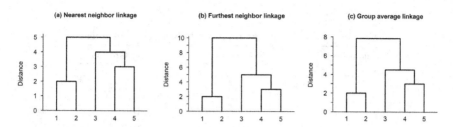

FIGURE 9.1

(a–c) Examples of dendrograms from cluster analyses of five objects.

There are many variations on the algorithm used, but the basic approach involves first choosing some more or less arbitrary group centers, with objects then allocated to the nearest one. New centers are then calculated, where these represent the averages of the objects in the groups. An object is moved to a new group if it is closer to that group's center than it is to the center of its present group. Any groups that are close together are merged, spread-out groups are split, and so on, following some defined rules. The process continues iteratively till stability is achieved with a predetermined number of groups. A range of values is usually tried for the final number of groups.

9.3 Hierarchical Methods

Agglomerative hierarchical methods start with a matrix of distances between objects. All objects begin alone in groups of size one, and groups that are close together are merged. There are various ways to define close. The simplest is in terms of nearest neighbors. For example, suppose that the distances between five objects are as shown in Table 9.1. The calculations are then as shown in Table 9.2.

TABLE 9.1

A Matrix Showing the Distances between Five Objects

	Object				
Object	1	2	3	4	5
1	—				
2	2	—			
3	6	5	—		
4	10	9	4	—	
5	9	8	5	3	—

Note: The distance is always zero between an object and itself, and the distance from object *I* to object *j* is the same as the distance from object *j* to object *i*.

TABLE 9.2

Merging of Groups Based on Nearest-Neighbor Distances

Distance	Groups
0	1, 2, 3, 4, 5
2	(1, 2), 3, 4, 5
3	(1, 2), 3, (4, 5)
4	(1, 2), (3, 4, 5)
5	(1, 2, 3, 4, 5)

TABLE 9.3

Merging of Groups Based on Furthest-Neighbor Distances

Distance	Groups
0	1, 2, 3, 4, 5
2	(1, 2), 3, 4, 5
3	(1, 2), 3, (4, 5)
5	(1, 2), (3, 4, 5)
10	(1, 2, 3, 4, 5)

Groups are merged at a given level of distance if one of the objects in one group that close or closer to at least one object in the second group. At a cut-off distance of 0, all five objects are on their own. The smallest distance between two objects is 2, which is between the first and second objects. Hence, at a distance level of 2, there are four groups: (1, 2), (3), (4), and (5). The next smallest distance between objects is 3, which is between objects 4 and 5. Hence, at a distance of 3, there are three groups (1, 2), (3), and (4, 5). The next smallest distance is 4, which is between objects 3 and 4. Hence, at this level of distance, there are two groups: (1, 2) and (3, 4, 5). Finally, the next smallest distance is 5, which is between objects 2 and 3 and between objects 3 and 5. At this level, the two groups merge into the single group (1, 2, 3, 4, 5), and the analysis is complete.

The dendrogram shown in Figure 9.1a illustrates how the agglomeration takes place. With furthest-neighbor linkage, two groups merge only if the most distant members of the two groups are close enough. With the example data, this works as shown in Table 9.3.

Object 3 does not join with objects 4 and 5 till distance level 5 because this is the distance to object 3 from the further away of objects 4 and 5. The furthest-neighbor dendrogram is shown in Figure 9.1b.

With group average linkage, two groups merge if the average distance between them is small enough. With the example data, this gives the results shown in Table 9.4. For example, groups (1, 2) and (3, 4, 5) merge at distance

TABLE 9.4

Merging of Groups Based on Group-Average Distances

Distance	Groups
0	1, 2, 3, 4, 5
2	(1, 2), 3, 4, 5
3	(1, 2), 3, (4, 5)
4.5	(1, 2), (3, 4, 5)
7.8	(1, 2, 3, 4, 5)

level 7.8, as this is the average distance from objects 1 and 2 to objects 3, 4, and 5, the actual distances being 1–3, 6; 1–4, 10, 1–5, 9; 2–3, 5; 2–4, 9; 2–5, 8, with (6 + 10 + 9 + 5 + 9 + 8)/6 = 7.8. The dendrogram in this case is shown in Figure 9.1c.

Divisive hierarchic methods have been used less often than agglomerative ones. The objects are all put into one group initially, and then this is split into two groups by separating off the object that is furthest on average from the other objects. Objects from the main group are then moved to the new group if they are closer to this group than they are to the main group. Further subdivisions occur as the distance that is allowed between objects in the same group is reduced. Eventually, all objects are in groups of size one.

9.4 Problems with Cluster Analysis

It has already been mentioned that there are many algorithms for cluster analysis. However, there is no generally accepted best method. Unfortunately, different algorithms do not necessarily produce the same results on a given set of data, and there is usually rather a large subjective component in the assessment of the results from any particular method.

A fair test of any algorithm is to take a set of data with a known group structure and see whether the algorithm is able to reproduce this structure. It seems to be the case that this test only works in cases where the groups are very distinct. When there is a considerable overlap between the initial groups, a cluster analysis may produce a solution that is quite different from the true situation.

In some cases, difficulties will arise because of the shape of clusters. For example, suppose that there are two variables X_1 and X_2, and objects are plotted according to their values for these. Some possible patterns of points are illustrated in Figure 9.2. Case (a) is likely to be found by any reasonable algorithm, as is case (b). In case (c), some algorithms might well fail to detect two clusters because of the intermediate points. Most algorithms would have trouble handling cases like (d), (e), and (f).

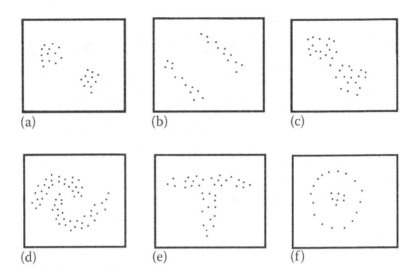

FIGURE 9.2
(a–f) Some possible patterns of points when there are two clusters.

Of course, clusters can only be based on the variables that are given in the data. Therefore, they must be relevant to the classification wanted. To classify depressed patients, there is presumably not much point in measuring height, weight, or length of arms. A problem here is that the clusters obtained may be rather sensitive to the particular choice of variables that is made. A different choice of variables, apparently equally reasonable, may give different clusters.

9.5 Measures of Distance

The data for a cluster analysis usually consist of the values of p variables $X_1, X_2, \ldots, X_p$ for n objects. For hierarchic algorithms, these values are then used to produce an array of distances between the objects. Measures of distance have already been discussed in Chapter 5. Here it suffices to say that the Euclidean distance function

$$d_{ij} = \left[\sum_{k=1}^{n} \left(x_{ik} - x_{jk} \right)^2 \right]^{1/2} \tag{9.1}$$

is often used, where x_{ik} is the value of variable X_k for object i and x_{jk} is the value of the same variable for object j.

The geometrical interpretation of the distance d_{ij} is illustrated in Figures 5.1 and 5.2 for the cases of two and three variables.

Usually, variables are standardized in some way before distances are calculated so that all p variables are equally important in determining these distances. This can be done by coding the variables so that the means are all zero and the variances are all one. Alternatively, each variable can be coded to have a minimum of zero and a maximum of one. Unfortunately, standardization has the effect of minimizing group differences because if groups are separated well by the variable X_k, then the variance of this variable will be large. In fact, it should be large. It would be best to be able to make the variances equal to one within clusters, but this is obviously not possible, as the whole point of the analysis is to find the clusters.

9.6 Principal Components Analysis with Cluster Analysis

Historically, some cluster analysis algorithms began by doing a principal component analysis to reduce a large number of original variables down to a smaller number of principal components. This can drastically reduce the computing time for the cluster analysis. However, the results of a cluster analysis can be rather different with and without the initial principal components analysis. Consequently, an initial principal components analysis is probably best avoided because computing time is seldom an issue currently.

On the other hand, when the first two principal components account for a high percentage of variation in the data, a plot of individuals against these two components is certainly a useful way for looking for clusters. For example, Figure 6.4 shows European countries plotted in this way for principal components based on how much they pay for a healthy diet. The countries do seem to group in a meaningful way.

9.6.1 Example 9.1: Clustering of European Countries

The data just mentioned on costs of six food types in different countries of Europe (Table 1.5) can be used for a first example of cluster analysis. The analysis should show which countries have similar cost patterns and which countries are different in this respect. As shown in Table 1.5, a grouping existed when the data were collected, consisting of (1) countries that joined the European Union until 1995 (EU95), (2) countries that joined the EU between 1996 and 2017 (EU17), (3) non-European Union countries (NEU), and (4) the two countries from the European Economic Area, Iceland, and Norway. It is, therefore, interesting to see whether this grouping can be recovered using a cluster analysis.

The first step in the analysis involves standardizing the six variables so that each one has a mean of zero and a standard deviation of one. For example, variable 1 is CSTS, the cost of starchy staples, measured as the purchasing power parity dollar per person per day, or PPP/day. For the 38 countries being considered, this variable has a mean of 0.358 PPP/day and a standard deviation of 0.124 PPP/day, with the latter value calculated using Equation 4.1. The CSTS data value for Austria is 0.229 PPP/day, which standardizes to (0.229–0.358)/0.124 = –1.04. Similarly, the CSTS data value for Belgium is 0.222 PPP/day, which standardizes to –1.10, and so on. The standardized data values are shown in Table 9.5.

The next step in the analysis involves calculating the Euclidean distances between all pairs of countries. This can be done by applying Equation 9.1 to the standardized data values. Finally, a dendrogram can be formed using, for example, the agglomerative, nearest-neighbor, hierarchic process described in Section 9.3. In practice, all these steps can be carried out using a suitable statistical package.

TABLE 9.5

Standardized Values for Costs of Six Food Types in Europe, Derived from the Costs in Table 5.1 Variable descriptions are given in Table 1.5.

Country	CSTS	CASF	CLNS	CVEG	CFRT	COFT
Albania	1.94	2.86	0.67	–0.15	1.45	0.23
Austria	–1.04	–0.81	–0.02	–0.20	0.76	–0.97
Belarus	1.19	0.52	2.39	–2.16	–0.30	3.21
Belgium	–1.10	0.01	0.06	–0.32	0.54	–0.79
Bosnia and Herzegovina	1.46	1.11	1.28	0.51	1.49	1.73
Bulgaria	1.17	1.37	0.99	0.22	1.81	0.65
Croatia	1.11	1.32	1.65	2.26	1.74	–0.34
Czechia	–0.49	–0.27	0.13	0.13	–0.19	–0.23
Denmark	–0.99	–1.52	–0.84	0.25	–1.29	–1.04
Estonia	–0.21	0.06	2.63	–1.08	0.18	0.08
Finland	–1.32	–1.24	0.38	–0.04	–0.91	–0.20
France	–0.91	–0.64	–0.72	0.64	1.06	–0.58
Germany	–0.99	–0.48	–0.52	0.46	–0.03	–1.07
Greece	0.41	0.34	0.41	0.04	–1.03	0.47
Hungary	0.39	0.39	0.49	0.11	0.83	0.23
Iceland	–0.66	–0.91	–0.72	–1.17	–1.75	–0.97
Ireland	–0.26	–1.36	–0.65	–0.85	–0.73	–0.79
Italy	–1.11	0.22	–0.82	0.09	0.71	–0.90

(Continued)

TABLE 9.5 (Continued)

Country	CSTS	CASF	CLNS	CVEG	CFRT	COFT
Latvia	−0.19	0.32	−0.41	1.04	−0.52	1.00
Lithuania	0.08	−0.25	−0.16	0.56	−0.36	0.12
Luxembourg	−1.21	−0.54	−0.70	−1.14	0.33	−0.93
Malta	1.20	0.32	0.75	−0.03	1.60	−0.09
Montenegro	0.83	0.02	−0.02	−2.21	−2.15	1.35
Netherlands	0.54	0.74	0.04	0.53	0.87	−0.06
North Macedonia	−0.29	−0.80	−0.91	0.88	−0.83	−1.04
Norway	1.34	1.28	−0.13	−0.13	−0.58	2.23
Poland	1.66	−0.32	−0.89	2.17	−0.77	−0.27
Portugal	−0.33	−0.99	0.13	0.87	−0.50	0.26
Republic of Moldova	−0.70	−0.99	−0.74	−0.64	−0.23	−0.16
Romania	0.00	1.26	−0.09	−1.33	−0.16	−0.34
Russian Federation	0.01	1.47	−1.45	0.03	0.21	1.84
Serbia	1.72	1.58	2.13	1.04	1.16	0.47
Slovakia	−0.07	−0.47	−0.43	1.34	−0.63	−0.20
Slovenia	−0.55	−0.28	−0.13	−0.07	−0.37	−0.37
Spain	−0.65	−0.74	−1.28	0.03	0.59	−1.04
Sweden	0.90	−0.61	−0.13	0.18	0.42	0.08
Switzerland	−1.18	−0.42	−0.96	0.05	−1.06	−0.55
United Kingdom	−1.70	−1.54	−1.42	−1.88	−1.39	−1.04

The dendrogram obtained using R (R Core Team, 2024) is shown in Figure 9.3. The two closest countries are Czechia and Slovenia. These are at about 0.40 units apart. At the larger distance of 0.78, Hungary joins Montenegro to make another cluster of size two. The next cluster merges the first one (Czechia and Slovenia) with Lithuania, at a distance of 0.87. As the distance increases, more and more countries combine, and the amalgamation ends with Belarus joining all the other countries in one cluster, at a distance of 3.5. One interpretation of the dendrogram is that there are just four clusters defined by a nearest-neighbor distance of about 2.5. These are (1) Belarus, (2) Moldova, (3) Estonia, and (4) all the other countries. Clusters (1), (2), and (3) are made of singletons that separates off two non-EU countries (Belarus and Moldova) and one EU17 country (Estonia) from everything else and suggests that the classification into EU95, EU17, EEA, and non-EU countries is not a good indicator of patterns of healthy food affordability. This

Country-Group

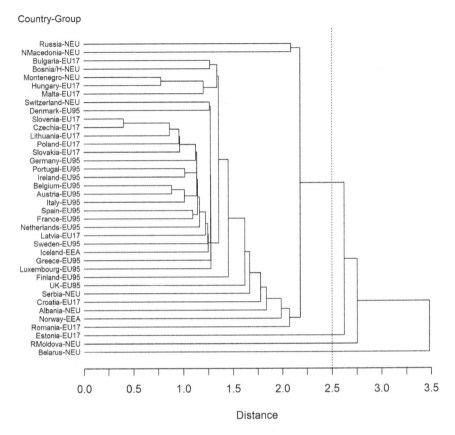

FIGURE 9.3
The dendrogram obtained from a nearest-neighbor, hierarchic cluster analysis on costs of
different food types from 38 European countries, originally arranged in four groups (EEA,
EU17, EU95, and NEU), as described in Table 1.5. The vertical dotted line highlights the nearest-
neighbor distance (2.5) used to define clusters. The line intersects three one-country clusters (or
singletons) and one large cluster composed by the remaining 35 countries.

contradicts the reasonably successful separation of non-EU and EU countries
by a discriminant function analysis (Figure 8.1). However, there is some limited agreement with the plot of countries against the first two principal components, where Belarus and Moldova show up as having very extreme data
values (Figure 6.4).

An alternative analysis was carried out using the K-means clustering option
in R. This essentially uses the partitioning method described in Section 9.2,
which starts with arbitrary cluster centers, allocates items to the nearest center, recalculates the mean values of variables for each group, reallocates individuals to their closest group centers to minimize the within-cluster total
sum of squares, and so on. The calculations use standardized variables with

means of zero and standard deviations of one. Ten random choices of start-ing clusters were tried, with from two to six clusters.

The percentage of the variation accounted for varied from 64.0% with two clusters to 31.9% with six clusters. With four clusters (and 42.2 % of variation accounted for), these were composed by (1) six countries, namely, three non-EU countries (Albania, Bosnia, and Herzegovina) and three EU17 countries (Bulgaria, Croatia, and Malta); (2) twelve EU95 countries (Austria, Belgium, Denmark, Finland, France, Germany, Ireland, Italy, Luxembourg, Netherlands, Portugal, Spain, and the UK), two EU17 countries (Czechia and Slovenia), one non-EU (Switzerland), and one EEA country (Iceland); (3) two EU95 countries (Greece and Sweden), six EU17 countries (Hungary, Latvia, Lithuania, Poland, Romania and Slovakia), three non-EU countries (Montenegro, Macedonia and Russia), and one EEA country (Sweden); and (4) the EU17 country of Estonia and the Non-EU countries of Belarus and Moldova. This grouping is not the same as the four-cluster solution given by the dendrogram of Figure 9.3. Indeed, the K-means algorithm put together in cluster 4 the three singleton clusters observed in the dendrogram shown in Figure 9.3, based on a nearest neighbor distance of 2.5. No doubt, other algo-rithms for cluster analysis will give slightly different solutions.

9.6.2 Example 9.2: Relationships between Canine Species

As a second example, consider the data provided in Table 1.4 for mean mandi-ble measurements of seven canine groups. As has been explained in Example 1.4 in Chapter 1, these data were originally collected as part of a study on the relationship between prehistoric dogs, whose remains have been uncovered in Thailand, and the other six living species. This question has already been considered in terms of distances between the seven groups in Example 5.1. Table 5.1 shows mandible measurements standardized to have means of zero and standard deviations of one. Table 5.2 shows Euclidean distances between the groups based on these standardized measurements.

With only seven species to cluster, it is a simple matter to carry out a near-est-neighbor, hierarchic cluster analysis without using a computer. Thus, it can be seen from Table 5.2 that the two most similar species are the prehis-toric dog and the modern dog, at a distance of 0.72. These, therefore, join into a single cluster at that level. The next largest distance is 1.38 between the cuon and the prehistoric dog so that at that level, the cuon joins the cluster with the prehistoric and modern dogs. The third largest distance is 1.63 between the cuon and the modern dog, but because these are already in the same clus-ter, this has no effect. Continuing in this way produces the clusters that are shown in Table 9.6. The corresponding dendrogram is shown in Figure 9.4.

It appears that the prehistoric dog is closely related to the modern Thai dog, with both being somewhat related to the cuon and dingo and less closely related to the golden jackal. The Indian and Chinese wolves are clos-est to each other, but the difference between them is relatively large. It seems

TABLE 9.6

Clusters Found at Different Distance Levels for a Hierarchic Nearest-Neighbor Cluster Analysis

Distance	Clusters	Number of clusters
0.00	MD, PD, GJ, CW, IW, CU, DI	7
0.72	(MD, PD), GJ, CW, IW, CU, DI	6
1.38	(MD, PD, CU), GJ, CW, IW, DI	5
1.63	(MD, PD, CU), GJ, CW, IW, DI	5
1.68	(MD, PD, CU, DI), GJ, CW, IW	4
1.80	(MD, PD, CU, DI), GJ, CW, IW	4
1.84	(MD, PD, CU, DI), GJ, CW, IW	4
2.07	(MD, PD, CU, DI, GJ), CW, IW	3
2.31	(MD, PD, CU, DI, GJ), (CW, IW)	2

Note: MD = modern dog, GJ = golden jackal, CW = Chinese wolf, IW = Indian wolf, CU = cuon, DI = dingo, and PD = prehistoric dog

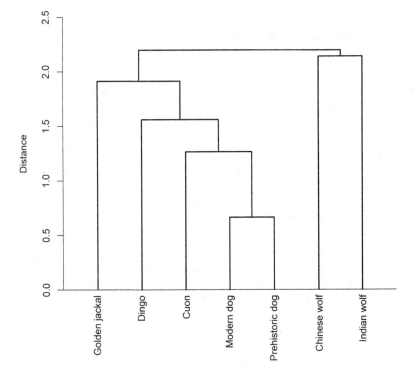

FIGURE 9.4
Dendrogram produced from a nearest-neighbor cluster analysis for the relationship between canine species.

fair to say that in this example, the cluster analysis has produced a sensible description of the relationship between the different groups.

9.7 Computer Programs

Computer programs for cluster analysis are widely available, and the larger statistical packages often include a variety of different options for both hierarchic and partitioning methods. As the results obtained usually vary to some extent depending on the precise details of the algorithms used, it will usually be worthwhile to try several options before deciding on the final method to be used for an analysis. Doing the calculations using methods available in R packages is described in the appendix to this chapter.

9.8 Discussion and Further Reading

A number of books devoted to cluster analysis are available, including the classic texts by Hartigan (1975), Kaufman and Rousseeuw (1990), Romesburg (2004), and Everitt et al. (2011). The book by Giordani et al. (2020) does a nice job of explaining how to use R for cluster analysis. An approach to clustering that has not been considered in this chapter involves assuming that the data available come from a mixture of several different populations for which the distributions are of a type that is assumed to be known (e.g., multivariate normal). The clustering problem is then transformed into the problem of estimating, for each of the populations, the parameters of the assumed distribution and the probability that an observation comes from that population. This has the merit of moving the clustering problem away from the development of ad hoc procedures toward the more usual statistical framework of parameter estimation and model testing. See Everitt et al. (2011, chapter 6) for an introduction to this method.

Exercises

1. Table 9.7 shows the abundances of the 25 most abundant plant species on 17 plots from a grazed meadow in Steneryd Nature Reserve in Sweden, as measured by Persson (1981) and used for an example by Digby and Kempton (1987). Each value in the table is the sum of cover values in a range from 0 to 5 for nine sample quadrats so that a value of 45 corresponds to complete cover by the species being considered. Note that the species are in order from the most abundant

(1) to the least abundant (25), and the plots are in the order given by Digby and Kempton (1987, table 3.2), which corresponds to variation in certain environmental factors such as light and moisture. Carry out cluster analyses to study the relationships between (a) the 17 plots and (b) the 25 species.

2. Table 9.8 partially shows data concerning grave goods from a cemetery at Bannadi, northeast Thailand, which were kindly supplied by Professor C.F.W. Higham. The full data **Bannadi_burials.csv**

TABLE 9.7

Abundance Measures for 25 Plant Species on 17 plots in Steneryd Nature Reserve, Sweden

Species	Plot																
	1	2	3	4	5	6	7	8	9	10	11	12	13	14	15	16	17
Festuca ovina	38	43	43	30	10	11	20	0	0	5	4	1	1	0	0	0	0
Anemone nemorosa	0	0	0	4	10	7	21	14	13	19	20	19	6	10	12	14	21
Stallaria holostea	0	0	0	0	0	6	8	21	39	31	7	12	0	16	11	6	9
Agrostis tenuis	10	12	19	15	16	9	0	9	28	8	0	4	0	0	0	0	0
Ranunculus ficaria	0	0	0	0	0	0	0	0	0	0	13	0	0	21	20	21	37
Mercurialis perennis	0	0	0	0	0	0	0	0	0	0	1	0	0	0	11	45	45
Poa pratenis	1	0	5	6	2	8	10	15	12	15	4	5	6	7	0	0	0
Rumex acetosa	0	7	0	10	9	9	3	9	8	9	2	5	5	1	7	0	0
Veronica chamaedrys	0	0	1	4	6	9	9	9	11	11	6	5	4	1	7	0	0
Dactylis glomerata	0	0	0	0	0	8	0	14	2	14	3	9	8	7	7	2	1
Fraxinus excelsior (juv.)	0	0	0	0	0	8	0	0	6	5	4	7	9	8	8	7	6
Saxifraga granulate	0	5	3	9	12	9	0	1	7	4	5	1	1	1	3	0	0
Deschampsia flexuosa	0	0	0	0	0	0	30	0	14	3	8	0	3	3	0	0	0
Luzula campestris	4	10	10	9	7	6	9	0	0	2	1	0	2	0	1	0	0
Plantago lanceolata	2	9	7	15	13	8	0	0	0	0	0	0	0	0	0	0	0
Festuca rubra	0	0	0	0	15	6	0	18	1	9	0	0	2	0	0	0	0
Hieracium pilosella	12	7	16	8	1	6	0	0	0	0	0	0	0	0	0	0	0
Geum urbanum	0	0	0	0	0	7	0	2	2	1	0	7	9	2	3	8	7
Lathyrus montanus	0	0	0	0	0	7	9	2	12	6	3	8	0	0	0	0	0
Campanula persicifolia	0	0	0	0	2	6	3	0	6	5	3	9	3	2	7	0	0
Viola riviniana	0	0	0	0	0	4	1	4	2	9	6	8	4	1	6	0	0
Hepatica nobilis	0	0	0	0	0	8	0	4	0	6	2	10	6	0	2	7	0
Achillea millefolium	1	9	16	9	5	2	0	0	0	0	0	0	0	0	0	0	0
Allium sp.	0	0	0	0	2	7	0	1	0	3	1	6	8	2	0	7	4
Trifolim repens	0	0	6	14	19	2	0	0	0	0	0	0	0	0	0	0	0

TABLE 9.8

Grave Goods in Burials in the Bannadi Cemetery in Northern Thailand The full data consists of 38 types of articles and 47 burials.

Burial	Type	1	2	3	...	36	37	38	Sum
				Types of articles					
B33	C	0	0	0	...	0	0	0	0
B9	F	0	0	0	...	0	0	0	0
B32	F	0	0	0	...	0	0	0	0
⋮	⋮	⋮	⋮	⋮	⋱	⋮	⋮	⋮	⋮
B47	M	0	0	1	...	1	0	1	8
B18	F	0	0	0	...	0	1	1	9
B48	F	0	0	0	...	1	1	1	11
	Sum	1	1	1	...	15	16	18	

Note: Body Types: C, child; F, adult female; M, adult male.

is accessible from the book's website and consist of a record of the presence or absence of 38 different articles in each of 47 graves, with additional information on whether the body was of a child, an adult female, or an adult male. The burials are in the order of richness of the different types of goods (from 0 to 11), and the goods are in the order of the frequency of occurrence (from 1 to 18). Carry out a cluster analysis to study the relationships between the 47 burials. Is there any clustering in terms of the type of body?

References

Digby, P.G.N., and Kempton, R.A. (1987). *Multivariate Analysis of Ecological Communities*. London: Chapman and Hall.

Everitt, B., Landau, S., Leese, M., and Stahl, D. (2011). *Cluster Analysis*. 5th Edn. New York: Wiley.

Giordani, P., Ferraro, M.B., and Martella, F. (2020). *An Introduction to Clustering with R*. Singapore: Springer.

Hartigan, J. (1975). *Clustering Algorithms*. New York: Wiley.

Kaufman, L., and Rousseeuw, P.J. (1990). *Finding Groups in Data. An Introduction to Cluster Analysis*. Hoboken, NJ: Wiley.

Persson, S. (1981). Ecological indicator values as an aid in the interpretation of ordination diagrams. *Journal of Ecology* 69: 71–84.

R Core Team. (2024). *R: A Language and Environment for Statistical Computing*. Vienna: R Foundation for Statistical Computing. www.R-project.org/.

Romesburg, H.C. (2004). *Cluster Analysis for Researchers*. Morrisville: Lulu.com.

Appendix: Cluster Analysis in R

The information needed to produce agglomerative dendrograms for a set of objects is performed by hclust, an R function that implements the most common algorithms for hierarchical clustering. The first argument of hclust is a distance or dissimilarity matrix (of objects), like any of those generated by the function dist, as described in the appendix for Chapter 5. The default algorithm in hclust is the furthest-neighbor or complete linkage (method = complete). Other agglomerative methods available include those described in Chapter 9 (single or nearest-neighbor, and average linkage) and extra methods indicated in the help documentation of hclust. The average linkage method is also known by the acronym *UPGMA*, which stands for *u*nweighted *p*air *g*roup *m*ethod with *a*rithmetic mean. As an example, given a distance matrix d.mat, the clustering of objects by means of the group average linkage (UPGMA) will be written as

```
clus.upgma <- hclust(d.mat, method = "average")
```

The resulting object clus.upgma contains crucial information about the clustering process. That object belongs to the class "hclust," and it can be used as the main argument for plotting a dendrogram. Using

```
plot(clus.upgma)
```

will produce a dendrogram with labels for each object hanging exactly where a branch starts, not necessarily at 0 distance. The branches of dendrograms displayed in Chapter 9 all start at 0, a situation that is forced by assigning a negative number after the option hang =. For example,

```
plot(clus.upgma, hang = -1)
```

generates a dendrogram of hclust object clus.upgma with all the labels hanging down from 0. To gain further control of a dendrogram's appearance using the plot function, it is more convenient to convert an hclust object into a dendrogram object with the command as.dendrogram. The objects of class "dendrogram" allow to handle tree-like structures in a more flexible way than objects of class "hclust". As an example, the following command will display a horizontal dendrogram based on the contents of the hclust object generated, with all object labels starting next to the 0 distance:

```
plot(as.dendrogram(clus.upgma), horiz = TRUE)
```

The class of object specified in this `plot` function is "dendrogram" and forces dendrogram labels to hang down from 0 by default; here, the argument `hang` = is not required.

More options are available for `hclust`. One is the command `cutree`, which allows the user to cut a dendrogram (tree) into several groups by specifying either the desired number of groups or the cut heights, that is, the distance at which groups are identifiable. Another useful command is `rect.hclust`, which draws rectangles around the branches of a dendrogram highlighting the corresponding clusters. With this command, first the dendrogram is cut at a certain level, and then a rectangle is drawn around selected branches.

Even the functions offered in the default `stats` package for dendrogram plotting with objects of class "dendrogram", as those described here, have limitations. An improvement of basic dendrogram plotting has been implemented in the package `dendextend`. As the name suggests, this package extends the functionality of "dendrogram" objects to produce sophisticated tree-like structures. Figure 8.3 is an example of a dendrogram built with the help of functions in the `dendextend` package.

Several computational tools have been implemented both in the standard R and in special cluster analysis packages looking for patterns in large data sets, a process called *data mining* (Kassambara, 2017). Thus, extra clustering techniques have been put together in the more extensive package `cluster` (Maechler et al., 2023), which includes functions for hierarchical and nonhierarchical clustering and other methods, all of them with feminine names, such as `agnes`, `clara`, `diana`, `mona`, and `pam`. The genesis of most algorithms implemented in `cluster` can be found in the classic book by Kaufman and Rousseeuw (1990). As an example, the power of `hclust` has been improved with `agnes`, one of the functions in the `cluster` package providing extra hierarchical clustering algorithms as well as an agglomerative coefficient measuring the amount of clustering structure found. It also provides a `banner`, which is a graphical display equivalent to a dendrogram. See the details in the help documentation of the `cluster` package. More advanced statistics practitioners and R users may also be interested to explore the `mclust` package (Scrucca et al., 2023) containing a variety of functions for model-based clustering, a topic briefly accounted in Section 9.8.

R scripts found at the book's website have been written to produce the dendrograms shown in Sections 9.3 (Figure 9.1) and 9.6 (Figures 9.3 and 9.4) using `hclust` and `agnes` functions. Another R script has also been included to show the application of `kmeans`, a function in the `stats` package useful to get the results of the nonhierarchical clustering method described in Section 9.2, for the European food data analyzed in Example 9.1.

References

Kassambara, A. (2017). *Practical Guide to Cluster Analysis in R. Unsupervised Machine Learning*. Marseille: STHDA. www.sthda.com.

Kaufman, L., and Rousseeuw, P.J. (1990). *Finding Groups in Data. An Introduction to Cluster Analysis*. Hoboken, NJ: Wiley.

Maechler, M., Rousseeuw, P., Struyf, A., Hubert, M., and Hornik, K. (2023). *Cluster: Cluster Analysis Basics and Extensions*. R package version 2.1.6. https://CRAN.R-project.org/package=cluster

Scrucca, L., Fraley, C., Murphy, T.B., and Raftery, A.E. (2023). *Model-Based Clustering, Classification, and Density Estimation Using mclust in R*. Boca Raton, FL: Chapman and Hall/CRC.

10

Canonical Correlation Analysis

10.1 Generalizing a Multiple Regression Analysis

For sets of multivariate data, the variables that divide naturally into two groups, a canonical correlation analysis can be used to investigate the relationships between the groups. A case in point is provided in Table 1.3 in which 16 colonies of the butterfly *Euphydryas editha* in California and Oregon are considered. For each colony, we have values for four environmental variables and six gene frequencies. What relationships, if any, exist between the gene frequencies and the environmental variables? One method to investigate this is a canonical correlation analysis.

Hotelling (1936) described a canonical correlation analysis for the first time. This example involved the results of tests for reading speed (X_1), reading power (X_2), arithmetic speed (Y_1), and arithmetic power (Y_2) for 140 seventh-grade schoolchildren. The specific question that was addressed was whether or not reading ability (as measured by X_1 and X_2) is related to arithmetic ability (as measured by Y_1 and Y_2).

The approach that a canonical correlation analysis takes to answering this question is to search for a linear combination of X_1 and X_2

$$U = a_1 X_1 + a_2 X_2$$

and a linear combination of Y_1 and Y_2

$$V = b_1 Y_1 + b_2 Y_2$$

where these are chosen to make the correlation between U and V as large as possible. This is similar to the idea behind a principal components analysis, except that here a correlation is maximized instead of a variance.

With the variables standardized to have unit variances, Hotelling found that the best choices for U and V with the reading and arithmetic example were $U = -2.78X_1 + 2.27X_2$ and $V = -2.44Y_1 + 1.00Y_2$, where these two variables have a correlation of 0.62. Thus, U measures the difference between

DOI: 10.1201/9781003453482-10

reading power and speed, and V measures the difference between arithmetic power and speed. It appears that that children with a large difference between X_1 and X_2 also tended to have a large difference between Y_1 and Y_2. It is this aspect of reading and arithmetic that shows the most correlation.

In a multiple regression analysis, a single variable Y is related to two or more variables $X_1, X_2, ...X_p$, to see how Y is related to the X variables. From this point of view, canonical correlation analysis is a generalization of multiple regression in which several Y variables are simultaneously related to several X variables.

In practice, more than one pair of canonical variables can be calculated from a set of data. Given $X_1, X_2, ...X_p$, and $Y_1, Y_2, ...Y_q$, then there can be up to the minimum of p and q pairs of variables. That is to say, linear relationships

$$U_i = \sum_{j=1}^{p} a_{ij} X_j; i = 1, 2, ..., r, \text{ and } V_i = \sum_{j=1}^{q} b_{ij} Y_j; i = 1, 2, ..., r,$$

can be established, where r is the smaller of p and q. These relationships are chosen so that the correlation between U_1 and V_1 is a maximum; then that between U_2 and V_2 is a maximum, subject to these variables being uncorrelated with the first two; the pattern continues. Each of the pairs of canonical variables $\left((U_1, V_1), (U_2, V_2), ..., (U_r, V_r)\right)$ then represents an independent dimension in the relationship between the two sets of original variables. The first new pair has the highest possible correlation and is, therefore, the most important, the second pair has the second highest correlation and is, therefore, the second most important, and so on.

10.2 Procedure for a Canonical Correlation Analysis

Let $\mathbf{R}_{XX}$ be the $p \times p$ matrix of correlations among the X variables, $\mathbf{R}_{YY}$ be the corresponding $q \times q$ matrix for the Y variables, and $\mathbf{R}_{XY}$ be the $p \times q$ correlation matrix between the X and the Ys. From these, a $q \times q$ matrix $\mathbf{R}_{YY}^{-1} \mathbf{R}_{XY}' \mathbf{R}_{XX}^{-1} \mathbf{R}_{XY}$ can be calculated, and the eigenvalue problem

$$\left(\mathbf{R}_{YY}^{-1} \mathbf{R}_{XY}' \mathbf{R}_{XX}^{-1} \mathbf{R}_{XY} - \lambda \mathbf{I}\right) \mathbf{b} = 0 \tag{10.1}$$

can be considered. It turns out that the eigenvalues $\lambda_1 > \lambda_2 > ... > \lambda_r$ are the squares of the correlations between the canonical variables, and the corresponding eigenvectors $\mathbf{b}_1, \mathbf{b}_2, ..., \mathbf{b}_r$ give the coefficients of the Y variables for the canonical variables. Recall that U_i is the ith canonical variable for the Xs; its coefficients are given by the elements of the vector

$$\mathbf{a}_i = \mathbf{R}_{YY}^{-1}\mathbf{R}'_{XY}\mathbf{b}_i \qquad (10.2).$$

In these calculations it is assumed that the original X and Y variables have been standardized to have means of zero and standard deviations of unity. The coefficients of the canonical variables are for these standardized variables.

From Equations 10.1 and 10.2, the ith pair of canonical variables are calculated as $U_i = \mathbf{a}'_i\mathbf{X}$ and $V_i = \mathbf{b}'_i\mathbf{Y}$, where $\mathbf{a}'_i = (a_{i1}, a_{i2}, \ldots, a_{ip})$, $\mathbf{b}'_i = (b_{i1}, b_{i2}, \ldots, b_{iq})$, $\mathbf{X}' = (x_1, x_2, \ldots, x_p)$ and $\mathbf{Y}' = (y_1, y_2, \ldots, y_q)$, with the X and Y values standardized. As they stand, U_i and V_i will have variances that depend on the scaling adopted for the eigenvector $\mathbf{b}_i$. However, it is a simple matter to rescale them to have unit variance. This standardization is not essential because the correlation between U_i and V_i is not affected by scaling. However, it may be useful when it comes to examining the numerical values of canonical variables for the individuals for which data are available.

10.3 Tests of Significance

An approximate test for a relationship between the X and Y variables, each taken as a whole, was proposed by Bartlett (1939, 1947) for data that are a random sample from a multivariate normal distribution. This involves calculating the following statistic:

$$X^2 = -\left(n - \frac{p+q+3}{2}\right)\sum_{i=1}^{r}\log_e(1 - \lambda_i) \qquad (10.3)$$

where n is the number of cases for which data are available. The statistic can be compared with the percentage points of the chi-squared (χ^2) distribution with pq degrees of freedom (df), and a significantly large value provides evidence that at least one of the r canonical correlations is significant. A non-significant result indicates that even the largest canonical correlation can be accounted for by sampling variation only.

It is sometimes suggested that this test can be extended to allow the importance of each of the canonical correlations to be tested. Common suggestions are the following:

1. Test the ith contributions of Equation 10.3 with the percentage points using a χ^2 distribution with $p + q - 2i + 1$ df.
2. Test the sum of the (i +1)th to the rth contributions of Equation 10.3 using a χ^2 distribution with $(p + i)(q - i)$ df.

The first approach tests the ith canonical correlation directly, whereas the second tests for the significance of the $(i + 1)$th to the rth canonical correlations as a whole.

The reason why these tests are not always reliable is essentially the same as has already been discussed in Section 8.4 for a related test used with discriminant function analysis. The ith largest sample canonical correlation in the sample may randomly not be the ith largest in the population. Hence, the association between the r contributions to Equation 10.3 and the r population canonical correlations is blurred. See Harris (2013) for further discussion of this matter.

There are also some modifications of the test statistic X^2 that are sometimes proposed to improve the chi-squared approximation for the distribution of this statistic when the null hypothesis holds and the sample size is small, but these will not be considered here.

10.4 Interpreting Canonical Variates

If

$$U_i = \sum_{j=1}^{p} a_{ij} X_j \text{ and } V_i = \sum_{j=1}^{q} b_{ij} Y_j,$$

then U_i can be interpreted in terms of the X variables with large coefficients a_{ij}, and V_i by the largest coefficients b_{ij}. Of course, *large* here means large in magnitude (positive or negative).

Unfortunately, correlations between the X and Y variables can upset this interpretation process. For example, a_{i1} may be positive, and yet the simple correlation between U_i and X_1 is negative. This apparent contradiction can come about if X_1 is highly correlated with one or more of the other X variables, and part of the effect of X_1 is being accounted for by the coefficients of these other X variables. In fact, if one of the X variables is very highly correlated with other X variables, then there will be an infinity of linear combinations of the X variables, some of them with very different a_{ij} values, that give virtually the same U_1 values. The same can be said with regard to the Y variables.

The interpretation problems that arise with highly correlated X or Y variables should be familiar to users of multiple regression analysis., since exactly the same problems arise with the estimation of regression coefficients.

Actually, if the X or Y variables are highly correlated, it might not be possible to disentangle their contributions to canonical variables, despite stout efforts to the contrary.

Some authors have suggested that it is better to describe canonical variables by looking at their correlations with the X and Y variables rather than the coefficients. For example, if U_i is highly positively correlated with X_1, then it can be considered to reflect X_1 to a large extent. Similarly, if V_i is highly negatively correlated with Y_1, then it can be considered to reflect the opposite of Y_1 to a large extent. This approach at least has the merit of bringing out all the variables to which the canonical variables seem to be related.

10.4.1 Example 10.1: Environmental and Genetic Correlations for Colonies of a Butterfly

The data in Table 1.3 can be used to illustrate the procedure for a canonical correlation analysis. Here, there are 16 colonies of the butterfly *Euphydryas editha* in California and Oregon. These vary with respect to four environmental variables (altitude, annual precipitation, annual maximum and minimum temperature), and six genetic variables (percentages of six phospho- glucose-isomerase genes as determined by electrophoresis). Any significant relationships between the environmental and genetic variables may indicate the adaption of *E. editha* to local environments.

For this analysis, the environmental variables have been treated as the X variables and the gene frequencies as the Y variables. However, not all the six gene frequencies have been used, as shown in Table 1.3, because they add up to 100%, which allows different linear combinations of these variables to have the same correlation with a combination of the X variables.

This problem is overcome by removing one of the gene frequencies from the analysis. Here, the 1.30 gene frequency was omitted. The data were also further modified by combining the low frequencies for the 0.40 and 0.60 mobility genes. Thus, the X variables remain the same, while the Y variables are Y_1: frequency of 0.40 and 0.60 mobility genes, Y_2: frequency of 0.80 mobility genes, Y_3: frequency of 1.00 mobility genes, and Y_4: frequency of 1.16 mobility genes. Following the development in Section 10.2, for the remainder of this example, we will use the standardized values of the variables.

The correlation matrix for the eight variables is shown in Table 10.1.

The eigenvalues obtained from Equation 10.1 are 0.743, 0.205, 0.143, and 0.007. Taking square roots gives the corresponding canonical correlations of 0.86, 0.45, 0.38, and 0.08, respectively, and the canonical variables are found to be:

$$U_1 = -0.12X_1 - 0.29X_2 + 0.47X_3 + 0.26X_4$$
$$V_1 = 0.55Y_1 + 0.42Y_2 - 0.09Y_3 + 0.83Y_4$$

$$U_2 = 2.43X_1 - 0.68X_2 + 0.48X_3 + 1.40X_4$$
$$V_2 = -1.77Y_1 - 2.26Y_2 - 3.85Y_3 - 2.85Y_4$$

TABLE 10.1

The Correlation Matrix for Variables Measured on Colonies of *Euphydryas editha*, Partitioned into the Three Matrices R_{XX}, R_{YY}, R_{XY}

	X_1	X_2	X_3	X_4	Y_1	Y_2	Y_3	Y_4
X_1	1.000	0.568	-0.828	-0.936	-0.201	-0.573	0.727	-0.458
X_2	0.568	1.000	-0.479	-0.705	-0.468	-0.550	0.699	-0.138
X_3 $\mathbf{R}_{XX}=$	-0.828	-0.479	1.000	0.719	0.224	0.536	-0.717	0.438
X_4	-0.936	-0.705	0.719	1.000	0.246	0.593	-0.759	0.412
Y_1	-0.201	-0.468	0.224	0.246	1.000	0.638	-0.561	-0.584
Y_2	-0.573	-0.550	0.536	0.593	0.638	1.000	-0.824	-0.127
Y_3 $\mathbf{R}_{XY}'=$	0.727	0.699	-0.717	-0.759	-0.561	-0.824	1.000	-0.264
Y_4	-0.458	-0.138	0.438	0.412	-0.584	-0.127	-0.264	1.000

$\mathbf{R}_{XY}=$ $\mathbf{R}_{YY}=$

$$U_3 = 2.95X_1 + 1.36X_2 + 0.58X_3 + 3.53X_4$$
$$V_3 = -3.48Y_1 - 1.30Y_2 - 3.75Y_3 - 2.75Y_4$$

$$U_4 = 1.37X_1 + 0.24X_2 + 1.70X_3 - 0.09X_4$$
$$V_4 = 0.66Y_1 - 1.41Y_2 - 0.50Y_3 + 0.64Y_4$$

There are four canonical correlations because this is the minimum of p and q, which here are both equal to four.

Note that some statistical packages may give one or more of the equations with opposite signs, for example with $U_1 = 0.12X_1 + 0.29X_2 - 0.47X_3 - 0.26X_4$. This would reverse the meaning of U_1 but not its usefulness for describing the data. The decision of swapping signs here is analogous to the case of a two-sample t-test for the difference in means between two groups (A and B, say). It is an arbitrary decision on your part whether to pursue the analysis using A − B or do it using B − A. It is common practice to choose the order of subtraction so as to yield an answer that doesn't have a negative sign (i.e., subtract the smaller mean from the larger). If by chance your statistics package produces a result with a negative sign, by switching the sign, all you are effectively doing is reversing the order of subtraction. The same statistical story is applied to multivariate statistics involving linear combinations with coefficients having particular signs. A suitable sign-swap decided by the data analyst perhaps tells an easier story.

Although the canonical correlations are quite large, they are not significantly so according to Bartlett's test because of the small sample size. It is found that $X^2 = 18.34$ with 16 df, where the probability of a value this large (or larger) from the chi-squared distribution is about 0.30.

Putting aside the lack of statistical significance, it is interesting to see what interpretation can be given to the first pair of canonical variables. The correlations between the environmental variables and U_1 are in the following table:

	Altitude	Precipitation	Maximum temperature	Minimum temperature
U_1	−0.92	−0.77	0.90	0.92

We can see that U_1 mainly a contrast between X_3 (maximum temperature) and X_4 (minimum temperature), on one hand, and X_2 (precipitation), on the other, suggesting that U_1 is best interpreted as a measure of high temperatures and low altitude and precipitation.

The correlations between V_1 and the gene frequencies are

	Mobility 0.40/0.60	Mobility 0.80	Mobility 1.00	Mobility 1.16
V_1	0.38	0.74	−0.96	0.48

For V_1, there are moderate to large positive coefficients for Y_1 (0.40 and 0.60 mobility), Y_2 (0.80 mobility), and Y_4 (1.16 mobility), and a small negative coefficient for Y_3 (1.00 mobility). It appears that the 0.40, 0.60, 0.80, and 1.16 mobility genes tend to be frequent in colonies with high temperatures and low precipitation. In this case, V_1 comes out clearly as indicating a lack of mobility 1.00 genes.

The interpretations of U_1 and V_1 are not the same when made on the basis of the coefficients of the canonical functions as they are on the basis of correlations. For U_1, the difference is not great and only concerns the status of altitude, but for V_1, the importance of the mobility 1.00 genes is very different. On the whole, the interpretations based on correlations seem best and correspond with what is seen in the data. For example, colony GL has the highest altitude, high precipitation, the lowest temperatures, and the highest frequency of 1.00 mobility genes. This compares with colony UO with a low altitude, low precipitation, high temperature, and the lowest frequency of mobility 1.00 genes. However, as mentioned in the previous section, there are real problems with interpreting canonical variables when the variables that they are constructed from have high correlations. Table 10.1 shows that this is indeed the case with this example.

Figure 10.1 shows a plot of the values of V_1 against the values of U_1. It is immediately clear that the colony labeled DP is somewhat unusual compared with the other colonies because the value of V_1 is different from that for other colonies with about the same values for U_1. From the interpretations given for U_1 and V_1, it seems that the frequency of mobility 1.00 genes is unusually high for a colony with this environment. Inspection of the data in Table 1.3 shows that this is the case.

10.4.2 Example 10.2: Soil and Vegetation Variables in Belize

For an example with a larger data set, Green (1973) studied the factors influencing the location of prehistoric Maya habitation sites in the Corozal district of Belize in Central America. Table 10.2 shows a subset for four soil variables and two vegetation variables recorded for $n = 151$ squares of sizes 2.5 × 2.5 km. The full data **Soil_and_Vegetation.csv** can be downloaded from the supplemental resource repository linked to the book. Here, canonical correlation analysis can be used to study the relationship between these two groups of variables.

The soil variables are X_1 = % soil with constant lime enrichment, X_2 = % meadow soil with calcium groundwater, X_3 = % soil with coral bedrock under

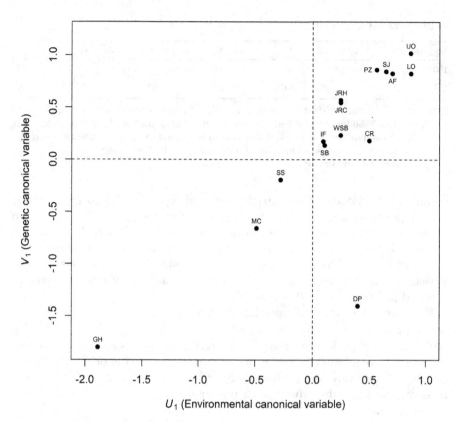

FIGURE 10.1

Plot of V_1 against U_1 for 16 colonies of *Euphydryas editha*.

TABLE 10.2

Example Soil and Vegetation Variables for 151 Plots in the Corozal Region of Belize

Square	X_1	X_2	X_3	X_4	Y_1	Y_2
1	40	30	0	30	0	25
2	20	0	0	10	10	90
3	5	0	0	50	20	50
4	30	0	0	30	0	60
5	40	20	0	20	0	9
⋮	⋮	⋮	⋮	⋮	⋮	⋮
147	0	0	100	0	75	25
148	90	0	10	0	60	30
149	0	0	100	0	80	10

(Continued)

TABLE 10.2   (Continued)

Square	X_1	X_2	X_3	X_4	Y_1	Y_2
150	0	0	100	0	60	40
151	0	40	60	50	50	50

Note: X_1 = % soil with constant lime enrichment, X_2 = % meadow soil with calcium groundwater, X_3 = % soil with coral bedrock under conditions of constant lime enrichment, and X_4 = % alluvial and organic soils adjacent to rivers and saline organic soil at the coast. Y_1 = % deciduous seasonal broadleaf forest, and Y_2 = % high and low marsh forest, herbaceous marsh, and swamp.

conditions of constant lime enrichment, and X_4 = % alluvial and organic soils adjacent to rivers and saline organic soil at the coast. The vegetation variables are Y_1 = % deciduous seasonal broadleaf forest and Y_2 = % marsh forest. The percentages do not add to 100, so there is no need to remove any variables before starting the analysis. It is the standardized values of these variables, with means of zero and standard deviations of one, that will be referred to for the rest of this example.

There are two canonical correlations (the minimum of p = 4 and q = 2), and they are found to be 0.686 and 0.481. From Equation 10.3, $X^2 = 131.70$, which is very significantly large. Therefore, there is very strong evidence that the soil and vegetation variables are related. However, the original data are clearly not normally distributed, so this result should be treated with some reservations. Here, the canonical correlations are

$$U_1 = 1.09X_1 + 0.07X_2 + 1.03X_3 + 0.23X_4$$
$$V_1 = 1.29Y_1 + 0.40Y_2$$

and

$$U_2 = 0.69X_1 + 1.02X_2 + 0.57X_3 + 0.88X_4$$
$$V_2 = 0.98Y_1 + 1.56Y_2.$$

By considering the correlations shown in Table 10.3 (particularly those large in magnitude), it appears that the canonical variables can be described as mainly measuring:

U_1: the presence of soil types 1 (soil with constant lime enrichment) and 3 (soil with coral bedrock under conditions of constant lime enrichment).

V_1: the presence of vegetation type 1 (deciduous seasonal broadleaf forest) and the absence of vegetation type 2 (high and low marsh forest, herbaceous marsh, and swamp).

TABLE 10.3

Correlations between the Canonical Variables and the X and Y Variables

	U_1	U_2		V_1	V_2
X_1	0.55	-0.10	Y_1	0.97	-0.25
X_2	-0.20	0.78	Y_2	-0.61	0.79
X_3	0.52	-0.02			
X_4	-0.53	0.33			

U_2: the presence of soil types 2 (meadow soil with calcium ground-water) and 4 (alluvial and organic soils adjacent to rivers and saline organic soil at the coast).

V_2: the presence of vegetation type 2 and the absence of vegetation type 1.

It appears, therefore, that the most important relationships between the soil and vegetation variables, as described by the first two pairs of canonical variables, are (a) the presence of soil types 1 and 3 and the absence of soil type 4 are associated with the presence of vegetation type 1, and (b) the presence of soil types 2 and 4 is associated with the presence of vegetation type 2 and the absence of vegetation type 1.

It is instructive to examine a plot of the canonical variables and the case numbers, as shown in Figure 10.2. The correlations between U_1 and V_1 and between U_2 and V_2 are apparent, as might be expected.

Before leaving this example, it is appropriate to mention a potential problem that has not been addressed. Squares that are close in space are spatially correlated, and particularly those that are adjacent. If such correlation exists so that, for example, neighboring squares tend to have the same soil and vegetation characteristics, then the data do not provide 151 independent observations. In effect, the data set will be equivalent to independent data from some smaller number of squares. The effect of this will appear mainly in the test for the significance of the canonical correlations as a whole, with a tendency for these correlations to appear to be more significant than they really are.

The same problem also potentially exists with the previous example on colonies of the butterfly *Euphydryas editha* because some of the colonies were quite close in space. Indeed, it is a potential problem whenever observations are taken in different locations in space. The way to avoid the problem is to ensure that observations are taken sufficiently far apart that they are independent or close to independent, although this is often easier said than done. There are methods available for allowing for spatial correlation in data, but these are beyond the scope of this book.

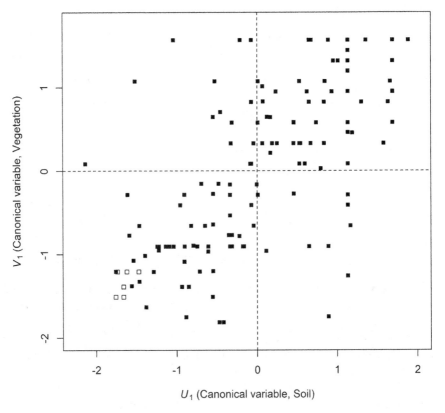

FIGURE 10.2
Plot of canonical variables obtained from the data on soil and vegetation variables for 2.5 km squares in Belize. Open squares indicate plots where soils of type 1 and 3 and vegetation type 1 are absent, while filled squares correspond to plots where at least one is nonzero.

10.5 Computer Programs

The appendix to this chapter provides information about R packages that can be used to carry out the analyses described in this chapter. However, the option for canonical correlation analysis is not as widely available in statistical packages as the options for the multivariate analyses considered in earlier chapters. Still, the larger packages certainly provide it.

10.6 Further Reading

There are not many books available that concentrate only on the theory and applications of canonical correlation analysis. A useful reference is the book by Giffins (1985) on applications of canonical correlation analysis in ecology. About half of this text is devoted to theory and the remainder to specific examples on plants. A shorter text with a social sciences emphasis is by Thompson (1984).

Exercise

Access the file **Cost_food_and_protein.csv** from the supplementary resources site. The first eight columns in this data set were considered in Example 1.5 (see Table 1.5), concerning to the costs of six food groups for 38 European countries. The last nine columns correspond to the availability of protein (g/capita/day) from nine food groups for the same countries: PCER = cereals and their products, PLNS = Nuts, seeds, and pulses, PFSH = fish and shellfish, PMLK = Milk and milk products, PEGG = Eggs and their products, PMEA = meat and meat products, PVEG = vegetables and their products, PFRT = Fruits and their products, POFT = Oil and fats. Use canonical correlation analysis to investigate the relationship, if any, between the nature of the economic accessibility of food groups and the amount of protein available.

References

Bartlett, M.S. (1939). A note on tests of significance in multivariate analysis. *Proceedings of the Cambridge Philosophical Society* 35: 180–185.

Bartlett, M.S. (1947). The general canonical correlation distribution. *Annals of Mathematical Statistics* 18: 1–17.

Giffins, R. (1985). *Canonical Analysis: A Review with Applications in Ecology*. Berlin: Springer.

Green, E.L. (1973). Location analysis of prehistoric Maya sites in British Honduras. *American Antiquity* 38: 279–293.

Harris, R.J. (2013). *A Primer of Multivariate Statistics*. 3rd Edn. New York and Hove: Psychology Press.

Hotelling, H. (1936). Relations between two sets of variables. *Biometrika* 28: 321–377.

Thompson, B. (1984). *Canonical Correlation Analysis: Uses and Interpretations*. Thousand Oaks, CA: SAGE.

Appendix: Canonical Correlation in R

Canonical correlation analysis is basically an eigenvalue problem, as defined by Equation 10.1, so it becomes evident that the canonical correlations for two sets of variables (say, X and Y variables) are easily computed in R with the function `eigen`. The programming effort is not remarkably reduced if the user decides to execute

```
cancor(matX, matY)
```

the basic command offered by R in the default package `stats` for canonical correlation analysis. Here, `matX` and `matY` are matrices containing the X and Y variables, respectively. `cancor` only allows data-centering, so `scale` must be run each time data standardization is required. The output of `cancor` is quite simple. It consists of the correlations, the coefficients of the linear combinations defining the canonical variables, and the means of the X and Y variables included in the analysis.

A wrapper of `cancor` (i.e., a function with the same name) has been included in `candisc` (Friendly and Fox, 2024), a package that was already considered in the appendix to Chapter 8 for discriminant function analysis. The function `cancor` implemented in `candisc` allows more general calculations and plots of the canonical variables in two dimensions. It also permits data standardization and provides a set of Wilks' lambda tests as alternatives to the Bartlett's tests described in Section 10.3. Moreover, the correlation matrices between the X and Y variables, and their corresponding canonical variables U_i and V_i, are part of the output produced by this improved version of `cancor`.

Bartlett's chi-squared test is included as the function `cca` found in `yacca` (Butts, 2022), a package whose acronym stands for "yet another package for canonical correlation analysis." The name `cca` is unfortunate here, as the data analyst may be confused by the same name being commonly used for the multivariate method *constrained correspondence analysis* (Oksanen, 2022). This method is mentioned in Chapter 12 with the name canonical correspondence analysis (Legendre and Legendre, 2012). A typical command involving `cca` for canonical correlation analysis takes the form

```
cca.object <- cca(matX, matY, xscale = TRUE, yscale = TRUE)
```

Here, `matX` and `matY` are the unstandardized data matrices whose columns are internally standardized by `cca` as a response to the options `xscale = TRUE` and `yscale = TRUE`.

There are three more packages in R containing functions for the standard canonical correlation analysis described in this chapter. One is the CCA package (González and Déjean, 2023), which includes the function `cc`, a regularized version of canonical correlation analysis to deal with data sets with more variables than units. An alternative package also handling with the issue of having more variables than units is `seedCCA` (Yoo and Kim, 2022; Kim et al., 2021), with the function `seedCCA`. The third package is `vegan` (Oksanen et al., 2022), in which a function called `CCorA` permits better canonical correlation analyses in cases of very sparse (with many zeros) and collinear matrices (with linearly dependent columns). The reader is encouraged to review the details of these three functions in the references given here and the R help documentation.

For the purposes of this primer, we recommend `cancor` implemented in `candisc`, and the function `cca` from `yacca` as the most suitable functions for canonical correlation analysis in R. In fact, both commands are sufficient to produce the results for Examples 10.1 and 10.2. The corresponding R scripts are available from the book's website.

References

Butts, C.T. (2022). *yacca: Yet Another Canonical Correlation Analysis Package*. R package version 1.4-2. https://CRAN.R-project.org/package=yacca.

Friendly, M., and Fox, J. (2024). *candisc: Visualizing Generalized Canonical Discriminant and Canonical Correlation Analysis*. R package version 0.9.0. http://CRAN.R-project.org/package=heplots.

González, I., and Déjean, S. (2023). *CCA: Canonical Correlation Analysis*. R package version 1.2.2. https://CRAN.R-project.org/package=CCA.

Kim, B., Im, Y., and Yoo, J.K. (2021). SEEDCCA: an integrated R-package for canonical correlation analysis and partial least squares. *The R Journal* 13(1): 7–20.

Legendre, P., and Legendre, L. (2012). *Numerical Ecology*. 3rd Edn. Amsterdam: Elsevier.

Oksanen, J. (2024). *Vegan: An Introduction to Ordination*. https://cran.r-project.org/web/packages/vegan/vignettes/intro-vegan.pdf.

Oksanen, J., Simpson, G., Blanchet, F., Kindt, R., Legendre, P., Minchin, P., O'Hara, R., Solymos, P., Stevens, M., Szoecs, E., Wagner, H., Barbour, M., Bedward, M., Bolker, B., Borcard, D., Carvalho, G., Chirico, M., De Caceres, M., Durand, S., Evangelista, H., FitzJohn, R., Friendly, M., Furneaux, B., Hannigan, G., Hill, M., Lahti, L., McGlinn, D., Ouellette, M., Ribeiro Cunha, E., Smith, T., Stier, A., Ter Braak, C., and Weedon, J. (2024). *vegan: Community Ecology Package*. R package version 2.6-6.1. https://CRAN.R-project.org/package=vegan.

Yoo, J.K., and Kim, B. (2022). *seedCCA: Seeded Canonical Correlation Analysis*. R package version 3.1. https://CRAN.R-project.org/package=seedCCA.

11

Multidimensional Scaling

11.1 Constructing a Map from a Distance Matrix

Multidimensional scaling is designed to construct a diagram showing the relationships between a number of objects, given only a table of distances between the objects. The diagram is a type of map, which can be in one dimension (if the objects fall on a line), in two dimensions (if the objects lie on a plane), in three dimensions (if the objects can be represented by points in space), or in a higher number of dimensions (in which case a simple geometrical representation is not possible).

We illustrate the construction of a map from a table of distances by considering objects A, B, C, and D (Figure 11.1; Table 11.1). For example, the distance from A to B (and vice versa) is 6.0; the distance from each object to itself is of course zero. The mirror image of that map (see Figure 11.2) is based on the same array of distances. This is much like the sign-switching of principal components and principal coordinates.

It is also apparent that if more than three objects are involved, then they may not lie on a plane. In that case, the distance matrix will implicitly contain this information. For example, the distance array shown in Table 11.2 requires three dimensions to show the spatial relationships between the four objects. Unfortunately, with real data, it is not usually known how many dimensions are needed for a representation. Hence, with real data, a range of dimensions usually must be tried.

The usefulness of multidimensional scaling comes from the fact that situations often arise where the underlying relationship between objects is not known, but a distance matrix can be estimated. For example, in psychology, subjects may be able to assess how similar or different individual pairs of objects are without being able to draw an overall picture of the relationships between the objects. Multidimensional scaling can then provide the picture.

At the present time, there are a wide variety of data analysis techniques that go under the general heading of multidimensional scaling. Here, only the simplest of these will be considered, these being the classical methods proposed by proposed by Togerson (1952) and Kruskal (1964a, 1964b). A related method called principal coordinates analysis is discussed in Chapter 12.

DOI: 10.1201/9781003453482-11

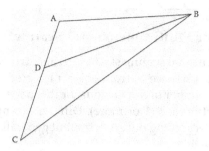

FIGURE 11.1
Four objects in two dimensions.

TABLE 11.1

Euclidean Distances between the Objects Shown in Figure 11.1

	A	B	C	D
A	0.0	6.0	6.0	2.5
B	6.0	0.0	9.5	7.8
C	6.0	9.5	0.0	3.5
D	2.5	7.8	3.5	0.0

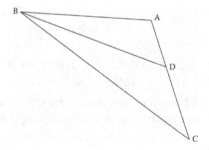

FIGURE 11.2
A mirror image of the objects in Figure 11.1, for which the distances between the objects are the same.

TABLE 11.2

A Matrix of Distances between Four Objects in Three Dimensions

	A	B	C	D
A	0	1	$\sqrt{2}$	$\sqrt{2}$
B	1	0	1	1
C	$\sqrt{2}$	1	0	$\sqrt{2}$
D	$\sqrt{2}$	1	$\sqrt{2}$	0

11.2 Procedure for Multidimensional Scaling

Classical multidimensional scaling starts with a matrix of distances between n objects that has the distance δ_{ij} from object i to object j, in the ith row and the jth column. The number of dimensions for the mapping of objects is fixed for a particular solution at t (1 or more). Different computer programs use different methods for carrying out analyses, but generally something like the following steps is involved.

1. A starting configuration is set up for the n objects in t dimensions; that is, coordinates $(x_1, x_2, ..., x_t)$ are assumed for each object in a t-dimensional space.

2. The Euclidean distances between the objects are calculated for the assumed configuration. Let d_{ij} be the distance between objects i and j for this configuration.

3. A regression of d_{ij} on δ_{ij} is made. The regression can be linear, polynomial, or monotonic. A monotonic regression assumes that if δ_{ij} increases, then d_{ij} either increases or remains constant, but no exact relationship between them is assumed. The fitted distances obtained from the regression equation (assuming a linear regression)

$$\hat{d} = \alpha + \beta\delta \qquad (11.1)$$

are called *disparities*. Thus, the disparities are the data distances scaled to match the configuration distances as closely as possible.

4. The goodness of fit between the configuration distances and the disparities is measured by a suitable statistic. One possibility is *Kruskal's stress formula 1*:

$$STRESS1 = \left(\sum\left(d_{ij} - \hat{d}_{ij}\right)^2 \middle/ \sum \hat{d}_{ij}^2\right)^{1/2} \qquad (11.2)$$

The word "stress" is used here because the statistic is a measure of the extent to which the spatial configuration of points has to be stressed in order to obtain the data distances.

5. The coordinates $(x_1, x_2, ..., x_t)$ of each object are changed slightly in such a way that the stress is reduced.

Steps 2 to 5 are repeated until it seems that the stress cannot be further reduced. The outcome of the analysis is then the coordinates of the n objects in t dimensions. These coordinates can be used to draw a map that shows how the objects are related. Hopefully a good solution is found in three or fewer dimensions, as a graphical representation of the n objects is then straightforward. Obviously, this is not always possible.

Small values of *STRESS1* (close to 0) are desirable. However, defining what is meant by "small" for a good solution is not straightforward. As a rough guide, Kruskal and Wish (1978, p. 56) indicate that reducing the number of dimensions to the extent that *STRESS1* exceeds 0.1, or increasing the number of dimensions when *STRESS1* is already less than 0.05, is questionable. However, their discussion concerning choosing the number of dimensions involves more considerations than this. In practice, the choice of the number of dimensions is often made subjectively, based on a compromise between the desire to keep the number small and the opposing desire to make the stress as small as possible. What is clear is that in general, there is little point in increasing the number of dimensions if this only leads to a small decrease in the stress.

An important distinction is between metric multidimensional scaling and nonmetric dimensional scaling. In the metric case, the configuration distances and the data distances are related by a linear or polynomial regression equation. With nonmetric scaling, all that is required is a monotonic regression, which means that only the ordering of the data distances is important. Generally, the greater flexibility of nonmetric scaling enables a better low-dimensional representation of the data to be obtained.

11.2.1 Example 11.1: Road Distances between New Zealand Towns

As an example of what can be achieved by multidimensional scaling, a map of the South Island of New Zealand has been constructed from a table of the road distances between the 13 towns shown in Figure 11.3.

If road distances were proportional to geographic distances, it would be possible to recover the true map exactly by a two-dimensional analysis. However, due to the absence of direct road links between many towns, road distances are in some cases far greater than geographic distances. Consequently, all that can be hoped for is an approximate recovery of the true map shown in Figure 11.3 from the road distances that are shown in Table 11.3.

The function metaMDS in the R package vegan (Oksanen et al., 2024) was used for the analysis. At Step 3 of the procedure described on page 04, a monotonic regression relationship was assumed between d_{ij} and δ_{ij}. This gives what is sometimes called a *classical nonmetric multidimensional scaling* or *classical NMDS*.

The program produced a two-dimensional solution for the data using the algorithm described. The final stress value was 0.036 as calculated using Equation 11.2.

The output from the program includes the coordinates of the 13 towns for the two dimensions produced in the analysis, as shown in Table 11.4. The

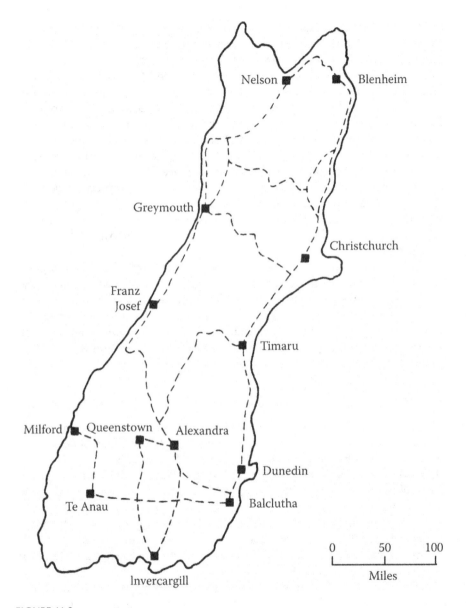

FIGURE 11.3
The South Island of New Zealand, with the main roads between 13 towns indicated by broken lines.

ordination dissimilarities are scaled to the same range as the input distances. To maintain the north–south and east–west orientation that exists between the real towns, the signs of the original values for both dimensions (NMDS1 and NMDS2) have been reversed to produce what are called *New Dimension*

TABLE 11.3

Main Road Distances in Miles between 13 Towns in the South Island of New Zealand

	Alexandra	Balclutha	Blenheim	Christchurch	Dunedin	Franz Josef	Greymouth	Invercargill	Milford	Nelson	Queenstown	Te Anau	Timaru
Alexandra	—												
Balclutha	100	—											
Blenheim	485	478	—										
Christchurch	284	276	201	—									
Dunedin	126	50	427	226	—								
Franz Josef	233	493	327	247	354	—							
Greymouth	347	402	214	158	352	114	—						
Invercargill	138	89	567	365	139	380	493	—					
Milford	248	213	691	489	263	416	555	174	—				
Nelson	563	537	73	267	493	300	187	632	756	—			
Queenstown	56	156	494	305	192	228	341	118	178	572	—		
Te Anau	173	138	615	414	188	366	480	99	75	681	117	—	
Timaru	197	177	300	99	127	313	225	266	377	366	230	315	—

TABLE 11.4

Coordinates Produced by Classical Nonmetric Multidimensional Scaling (NMDS) Applied to the Distances between Towns in the South Island of New Zealand

Town	Dimension	
	New 1 = –NMDS1	New 2 = –NMDS2
Alexandra	103.99	–42.84
Balclutha	150.49	100.75
Blenheim	–355.48	75.44
Christchurch	–148.00	66.87
Dunedin	103.29	115.25
Franz Josef	–130.24	–205.68
Greymouth	–220.23	–97.48
Invercargill	226.35	50.15
Milford	336.27	–64.10
Nelson	–416.98	0.38
Queenstown	137.95	–88.05
Te Anau	263.14	–22.73
Timaru	–50.56	112.03

Note: The signs of the original dimensions NMDS1 and NMDS2, produced by R, have been reversed to match the geographical locations of the real towns.

1 and *New Dimension 2*. These sign reversals do not change the distances between the towns based on the two dimensions, and the new dimensions are, therefore, just as satisfactory as the original ones. If the signs are left unchanged, then the plot of the towns against the two dimensions are placed in opposite positions of the real map.

A plot of the towns using these coordinates is shown in Figure 11.4. A comparison of this figure with Figure 11.3 indicates that the multidimensional scaling has been quite successful in recovering the real map. On the whole, the towns are shown with the correct relationships to each other. An exception is Milford. Because this town can only be reached by road through Te Anau, the map produced by multidimensional scaling has made Milford closest to Te Anau. In fact, Milford is geographically closer to Queenstown than it is to Te Anau.

To discern how well the *t*-dimensional model fits the data, multidimensional scaling plots are accompanied by a Shepard diagram, a scatterplot with the data dissimilarities on the horizontal axis and the fitted configuration distances or disparities on the vertical axis. A perfect representation of the data would show the configuration distances always increasing with the data distances. Computer programs usually add the best-fitting monotone step function through the Shepard plot, as a visual aid to detect outliers, and to discover individual dissimilarities not well represented by the

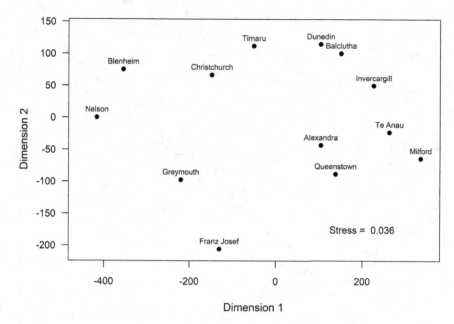

FIGURE 11.4
The map produced by a multidimensional scaling using the distances between New Zealand towns shown in Table 11.3.

multidimensional scaling solution (de Leeuw and Mair, 2015). The vertical distance between a point and the line indicates the error of the corresponding distance in the configuration generated by multidimensional scaling. High steps in that line indicate large discrepancies between data distances and their corresponding disparities. Figure 11.5 shows the Shepard diagram produced in R for the nonmetric multidimensional scaling solution in two dimensions of New Zealand Road distances. The largest discrepancies occur for towns around 500–600 miles away but, overall, it is confirmed that the fitted 2D model is satisfactory.

11.2.2 Example 11.2: The Voting Behavior of Congressmen

For a second example of the value of multidimensional scaling, consider the distance matrix shown in Table 11.5. Here, the distances are between 15 New Jersey congressmen in the United States House of Representatives. They are counts of the number of voting disagreements on 19 bills concerned with environmental matters. For example, Congressmen Hunt and Sandman disagreed 8 out of the 19 times, Sandman and Howard disagreed 17 out of the 19 times, and so on. An agreement was considered to occur if two congressmen both voted yes, both voted no, or both failed to vote. The table of distances was constructed from original data given by Romesburg (2004).

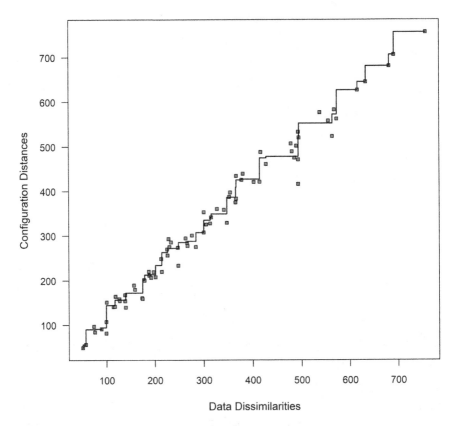

FIGURE 11.5
Shepard diagram for the nonmetric 2D solution of the road distances between New Zealand towns. The broken line through the points is the best-fitting monotone step function.

Two analyses were carried out in R (R Core Team, 2024). The first was a classical metric multidimensional scaling (Section 12.3) using the R-function `cmdscale` (see the appendix for Chapter 12) which assumes that the distances δ_{ij} of Table 11.5 are measured on a ratio scale (e.g., that a doubling of distance values implies a doubling of configuration distances). A simple linear regression at Step 3 was used. The stress values obtained for two-, three-, and four-dimensional solutions were found on this basis to be 0.237, 0.131, and 0.081, respectively.

The second analysis was carried out using the function `metaMDS` implemented in the R package `vegan` for a classical nonmetric scaling, so that the regression of d_{ij} on δ_{ij} was assumed to be monotonic only. In this case, the stress values for two-, three-, and four-dimensional solutions were found to be 0.066, 0.029, and 0.013, respectively. The distinctly lower stress values for nonmetric scaling suggest that this is preferable to metric scaling for these data, and the three-dimensional nonmetric solution has only slightly more

TABLE 11.5

The Distances between 15 Congressmen from New Jersey in the United States House of Representatives

	Hunt	Sandman	Howard	Thompson	Frelinghuysen	Forsythe	Widnall	Roe	Helstoski	Rodino	Minish	Rinaldo	Maraziti	Daniels	Pattern
Hunt (R)	0														
Sandman (R)	8	0													
Howard (D)	15	17	0												
Thompson (D)	15	12	9	0											
Frelinghuysen (R)	10	13	16	14	0										
Forsythe (R)	9	13	12	12	8	0									
Widnall (R)	7	12	15	13	9	7	0								
Roe (D)	15	16	5	10	13	12	17	0							
Helstoski (D)	16	17	5	8	14	11	16	4	0						
Rodino (D)	14	15	6	8	12	10	15	5	3	0					
Minish (D)	15	16	5	8	12	9	14	5	2	1	0				
Rinaldo (R)	16	17	4	6	12	10	15	3	1	2	1	0			
Maraziti (R)	7	13	11	15	10	6	10	12	13	11	12	12	0		
Daniels (D)	11	12	10	10	11	6	11	7	7	4	5	6	9	0	
Pattern (D)	13	16	7	7	11	10	13	6	5	5	5	4	13	9	0

Note: The numbers shown are the number of times that the congressmen voted differently on 19 environmental bills. R = Republican Party, D = Democratic Party.

TABLE 11.6

Coordinates of 15 Congressmen Obtained from a Three-Dimensional Nonmetric
Multidimensional Scaling Based on Voting Behavior

Congressman	Dimension		
	1	2	3
Hunt (R)	−8.97	1.82	−1.82
Sandman (R)	−8.06	8.22	0.65
Howard (D)	5.48	1.53	−3.23
Thompson (D)	3.33	4.95	2.54
Frelinghuysen (R)	−5.40	−4.69	5.33
Forsythe (R)	−3.35	−3.05	−0.83
Widnall (R)	−9.19	−2.12	1.05
Roe (D)	6.49	0.78	−0.72
Helstoski (D)	6.33	−0.82	0.18
Rodino (D)	4.12	−1.19	−0.46
Minish (D)	4.17	−1.30	−0.13
Rinaldo (R)	5.45	0.14	0.37
Maraziti (R)	−4.99	−2.74	−4.86
Daniels (D)	0.46	−1.38	−1.55
Pattern(D)	4.12	−0.13	3.49

Note: R = Republican Party, D = Democratic Party.

stress than the four-dimensional solution. This three-dimensional non-metric solution is, therefore, the one that will be considered in more detail. Table 11.6 shows the coordinates of the congressmen for the three-dimensional solution, and plots of the congressmen against the three dimensions are shown in Figure 11.6.

From Figure 11.6, it is clear that Dimension 1 is largely reflecting party differences, because the Democrats fall on the right-hand side of the figure, and the Republicans, other than Rinaldo, fall on the left-hand side.

To interpret Dimension 2, it is necessary to consider what it is about the voting of Sandman and Thompson, who have the highest two scores, that contrasts with Frelinghuysen, who has the lowest score. This points to the number of abstentions from voting. Sandman abstained from nine votes and Thompson abstained from six votes. On the other hand, Frelinghuysen abstained from three votes, but no other congressmen abstained voting on two of the bills where he did. The remaining individuals with low scores on Dimension 2 voted all or most of the time.

Dimension 3 appears to have no simple or obvious interpretation, although it must reflect certain aspects of differences in voting patterns. It suffices to

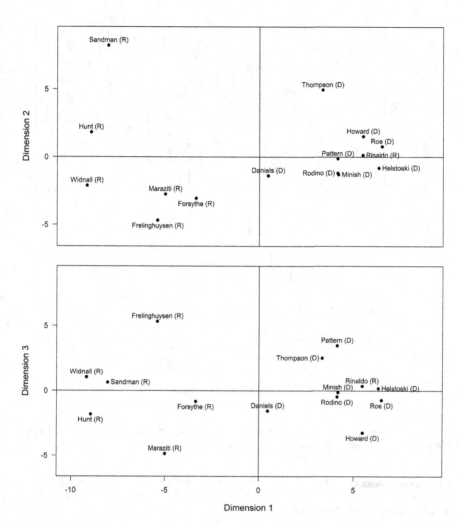

FIGURE 11.6
Plots of congressmen against the three dimensions obtained from a nonmetric multidimensional scaling.

say that the analysis has produced a representation of the congressmen in three dimensions that indicates how they relate with regard to voting on environmental issues.

Figure 11.7 shows the Shepard diagram of distances between the congressmen for the disparities against the original data. It is apparent that the three-dimensional model does not fit the data well. There is a wide range of configuration distances associated with each of the discrete data, with data distances of 10 showing up as the worse fitted by the model.

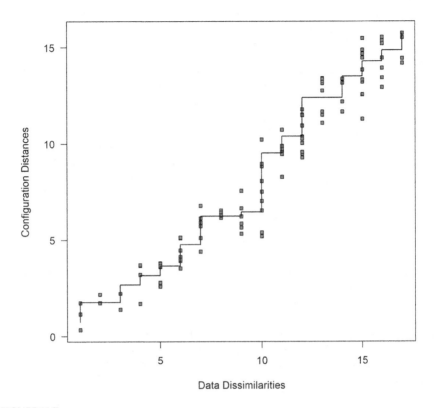

FIGURE 11.7
Shepard diagram of the congressmen fitted distances against the original data distances obtained for a three-dimensional configuration.

11.3 Computer Programs

The appendix to this chapter provides details on R packages that can be used for multidimensional scaling analyses. Some of the standard statistical packages also include a multidimensional scaling option, but in general, it can be expected that different packages may use slightly different algorithms and, therefore, may not give exactly the same results. Attention must be put to any convergence issues that might emerge, as the non-linear optimization algorithm implemented in the statistical package chosen may not reach the best possible solution, namely, the global minimum stress. Lack of convergence issues are reduced if multidimensional scaling is run with low dimension numbers ($t = 2$ or $t = 3$). If much higher number of dimensions is required, usually the statistical software suggests strategies to safeguard for convergence so, it is encouraged to carefully read the corresponding software manuals.

11.4 Further Reading

The classic book by Kruskal and Wish (1978) provides a short introduction to multidimensional scaling. More comprehensive treatments of the theory and applications of this topic and related topics are provided by Cox and Cox (2000) and Borg et al. (2013).

Exercise

Consider the data on costs of healthy food in 38 countries in Europe in Table 1.5. From these data, construct a matrix of Euclidean distances between the countries using Equation 5.1. Carry out nonmetric multidimensional scaling using this matrix to find out how many dimensions are needed to represent the countries in a manner that reflects differences in their economic access to healthy food.

References

Borg, I., Groenen, P.J.F., and Mair, P. (2013). *Applied Multidimensional Scaling*. Heidelberg: Springer.

Cox, T.F., and Cox, M.A.A. (2000). *Multidimensional Scaling*. 2nd Edn. Boca Raton, FL: Chapman and Hall/CRC.

de Leeuw, J., and Mair, P. (2015). Shepard diagram. In *Wiley StatsRef: Statistics Reference Online*. https://doi.org/10.1002/9781118445112.stat06268.pub2.

Kruskal, J.B. (1964a). Multidimensional scaling by optimizing goodness of fit to a nonmetric hypothesis. *Psychometrics* 29: 1–27.

Kruskal, J.B. (1964b). Nonmetric multidimensional scaling: a numerical method. *Psychometrics* 29: 115–129.

Kruskal, J.B., and Wish, M. (1978). *Multidimensional Scaling*. Thousand Oaks, CA: SAGE.

Oksanen, J., Simpson, G., Blanchet, F., Kindt, R., Legendre, P., Minchin, P., O'Hara, R., Solymos, P., Stevens, M., Szoecs, E., Wagner, H., Barbour, M., Bedward, M., Bolker, B., Borcard, D., Carvalho, G., Chirico, M., De Caceres, M., Durand, S., Evangelista, H., FitzJohn, R., Friendly, M., Furneaux, B., Hannigan, G., Hill, M., Lahti, L., McGlinn, D., Ouellette, M., Ribeiro Cunha, E., Smith, T., Stier, A., Ter Braak, C., and Weedon, J. (2024). *vegan: Community Ecology Package*. R package version 2.6-6.1. https://CRAN.R-project.org/package=vegan.

R Core Team. (2024). *R: A Language and Environment for Statistical Computing*. Vienna: R Foundation for Statistical Computing. www.R-project.org/.

Romesburg, H.C. (2004). *Cluster Analysis for Researchers*. Morrisville, NC: Lulu.com.

Togerson, W.S. (1952). Multidimensional scaling. 1. Theory and method. *Psychometrics* 17: 401–419.

Appendix: Multidimensional Scaling in R

When choosing the proper R command for multidimensional scaling, it is important to recall the difference between the classical metric and the nonmetric versions of this method. The classical multidimensional scaling is also known as principal coordinates analysis, a topic covered in Chapter 12. The standard acronym for this analysis is principal coordinate ordination (PCO). For this analysis, cmdscale is the main R function that has to be run, as described in the appendix for Chapter 12. The present chapter is concerned with Kruskal's nonmetric multidimensional scaling (NMDS). This has been implemented in isoMDS, a function in the MASS package (Venables and Ripley, 2002). The main argument of this function is a distance matrix that is a full, symmetric matrix or is created by the function dist. As an example, assuming that sym.mat is a symmetric matrix, the command

```
NMDS.obj <- isoMDS(sym.mat)
```

will store the results of the NMDS in the object NMDS.obj. The defaults assumed when running isoMDS with a distance matrix as the only argument are k (the number of dimensions, which is 2), maxit (the number of evaluations of stress [Equation 11.2] until convergence is 50), and tol (the tolerance is 1×10^{-3}; that is, once two consecutive values of the stress in the iterative process differ by 1×10^{-3} or less, the process stops, and an outcome for the analysis is produced). For a particular dimension chosen, sometimes the defaults for the number of iterations and the tolerance are somewhat slack, and a better multidimensional scaling configuration can be achieved if the user increases the number of iterations and/or decreases the tolerance. This is illustrated in the R script, accessible from the book's website, that has been written to carry out the analysis for the data in Example 11.1.

According to Step 1 in the procedure for NMDS given in Section 11.2, isoMDS uses an initial configuration of objects, either supplied by the user or generated by a PCO of the data, which is the default, with an automated intervention of the cmdscale function. The object NMDS.obj produced by isoMDS is a two-object list. The first is a scalar for the stress, namely, NMDS. obj$stress, given as percent. The second is the matrix NMDS.obj$points, which carries the coordinates of the sampling units in the reduced dimension chosen. Any pair of dimensions can be selected, and the corresponding coordinates displayed in a two-dimensional plot like Figures 11.4 and 11.6 using the plot command. Also, the original distance matrix, sym.mat, and the matrix NMDS.obj$points can be used as arguments of the function called Shepard, to produce the Shepard diagrams in Figures 11.5 and 11.7. See the R scripts for this chapter for details about the way to obtain this plot.

One alternative to `isoMDS` is offered by `metaMDS`, which is part of the `vegan` package (Oksanen et al., 2024). The developers of the `metaMDS` routine emphasize that it allows greater automation of the multidimensional scaling process than `isoMDS`. Also, `metaMDS` tries to eliminate the inaccuracies of `isoMDS` when this is run with the defaults and convergence is not guaranteed. Actually, `metaMDS` can use `isoMDS` in its calculations, but `metaMDS` is more versatile because it allows random starts of the function `initMDS` for the object configuration (Step 1 in Section 11.2), and scaling and rotation of the results (function `postMDS`) are available. This makes the routine more similar to the eigenvalue methods, as the final configuration of the NMDS is followed by a rotation via principal components analysis, so that, for example, the NMDS axis 1 reflects the principal source of variation.

The objects that `vegan` manipulates usually refer to terminology employed in ordination methods applied to community ecology. This explains the use of one specific parameter in `metaMDS` depending on whether community data is being analyzed or not. The following command invokes `metaMDS` to produce a nonmetric multidimensional scaling in k = 3 dimensions, when the data type does not consist of community data, as in Examples 11.1 and 11.2. In those cases, the `autotransform` parameter is set to FALSE:

```
MDSdim3 <- metaMDS(distmat, k = 3, autotransform = FALSE))
```

The first argument in this command refers to the input distance matrix `distmat`, either an object of `dist` class or a symmetric square matrix. The output object `MDSdim3` contains the most important statistics produced by the analysis, including the stress and the coordinates of the objects for the chosen number of dimensions. In the line of the default data types analyzed by `vegan`, `metaMDS` identifies objects with the name of `sites`, as these are usually the sampling units in community data. The following command allows to extract the coordinates (*scores*) of objects from the `metaMDS` object `MDSdim3`:

```
coords.3D <- scores(MDSdim3, display = "sites")
```

The `vegan` package also implements one function like the `Shepard` command described. The referred function is `stressplot`, and the only required argument for its execution is the name of the object generated by `metaMDS`:

```
stressplot(MDSdim3)
```

Other options and functions complementing the basic execution of `metaMDS` described here are found in the help files of `vegan`.

A third option in multidimensional scaling analysis uses the application of optimization (majorization) algorithms, where a particular objective function such as the stress has to be minimized. A collection of R commands following the principle of majorization has been put together in the `smacof` package for scaling by *majorizing* a *complicated function* (de Leeuw and Mair, 2009). In particular, with `smacofSym` or the wrapper function `mds`, it is possible to perform multidimensional scaling on any symmetric dissimilarity matrix. When the option `type = "ordinal"` is written as an argument of this function, a nonmetric multidimensional scaling algorithm is used. For details, see the paper by Mair et al. (2022) or the R documentation for the `smacof` package. Finally, the reader may like to run the R scripts that are accessible at the book's website containing the commands `metaMDS` and `mds` as alternative ways to produce the NMDS analyses illustrated in Examples 11.1 and 11.2.

References

de Leeuw, J., and Mair, P. (2009). Multidimensional scaling using majorization: SMACOF in R. *Journal of Statistical Software* 31: 1–30.

Mair, P., Groenen, P., and de Leeuw, J. (2022). More on multidimensional scaling and unfolding in R: smacof version 2. *Journal of Statistical Software* 102(10): 1–47. https://doi.org/10.18637/jss.v102.i10

Oksanen, J., Simpson, G., Blanchet, F., Kindt, R., Legendre, P., Minchin, P., O'Hara, R., Solymos, P., Stevens, M., Szoecs, E., Wagner, H., Barbour, M., Bedward, M., Bolker, B., Borcard, D., Carvalho, G., Chirico, M., De Caceres, M., Durand, S., Evangelista, H., FitzJohn, R., Friendly, M., Furneaux, B., Hannigan, G., Hill, M., Lahti, L., McGlinn, D., Ouellette, M., Ribeiro Cunha, E., Smith, T., Stier, A., Ter Braak, C., and Weedon, J. (2024). *vegan: Community Ecology Package*. R package version 2.6-6.1. https://CRAN.R-project.org/package=vegan.

Venables, W.N., and Ripley, B.D. (2002). *Modern Applied Statistics*. 4th Edn. New York: Springer.

12

Ordination

12.1 The Ordination Problem

The word "ordination" for a biologist means essentially the same as scaling does for a social scientist. Both words describe the process of producing a small number of variables that can be used to describe the relationship between a group of objects, starting either from a matrix of distances or similarities between the objects or from the values of some variables measured on each object. From this point of view, many of the methods that have been described in earlier chapters can be used for ordination, and some of the examples have been concerned with this process. In particular, plotting female sparrows against the first two principal components of size measurements (Example 5.1), plotting European countries against the first two principal components for the costs of food types (Example 5.2), producing a map of the South Island of New Zealand from a table of distances between towns by multidimensional scaling (Example 11.1), and plotting New Jersey congressmen against axes obtained by multidimensional scaling based on voting behavior (Example 11.2) are all examples of ordination. In addition, discriminant function analysis is a type of ordination designed to emphasize the differences between objects in different groups, while canonical correlation analysis is designed to emphasize the relationships between two groups of variables measured on the same objects.

Although ordination covers a diverse range of situations, in biology it is often used as a means of summarizing the relationships between different species as determined from their abundances at a number of different locations or, alternatively, as a means of summarizing the relationships between different locations on the basis of the abundances of different species at those locations. It is this type of application that is considered in this chapter, although the examples involve archaeology as well as biology. The purpose of the chapter is to give more examples of the use of principal components analysis and multidimensional scaling in this context, and to describe the methods of principal coordinates analysis and correspondence analysis that have not been covered in earlier chapters.

DOI: 10.1201/9781003453482-12

12.2 Principal Components Analysis

Principal components analysis has already been discussed in Chapter 6. Recall that it is a method whereby the values for variables $X_1, X_2, \ldots, X_p$ measured on each of n objects are used to construct principal components $Z_1, Z_2, \ldots, Z_p$ that are linear combinations of the X variables and are such that Z_1 has the maximum possible variance; Z_2 has the largest possible variance, conditional on it being uncorrelated with Z_1; Z_3 has the maximum possible variance, conditional on it being uncorrelated with both Z_1 and Z_2; and so on. The idea is that it may be possible for some purposes to replace the X variables with a smaller number of principal components, with little loss of information.

In terms of ordination, if, say, the first two principal components are sufficient to describe the differences between the objects because then a plot of Z_2 against Z_1 provides what is required. It is less satisfactory to find that three principal components are important, but a plot of Z_2 against Z_1 with values of Z_3 indicated may be acceptable. If four or more principal components are important, then easy graphical representation is not available.

12.2.1 Example 12.1: Plant Species in the Steneryd Nature Reserve

Table 9.7 shows the abundances of 25 plant species on 17 plots from a grazed meadow in Steneryd Nature Reserve in Sweden, as described in Exercise 1 of Chapter 9, which used the data for cluster analyses. Here an ordination of the plots will be considered so that the variables for principal components analysis are the abundances of the plant species. In other words, in Table 9.7 the objects of interest are the plots (columns), and the variables are the species (rows).

Because there are more species than plots, the number of nonzero eigenvalues in the correlation matrix is determined by the number of plots. In fact, there are 16 nonzero eigenvalues, as shown in Table 12.1. The first three components account for about 69% of the variation in the data, which is not a particularly high amount. The coefficients for the first three principal components are shown in Table 12.2. They are all contrasts between the abundance of different species that may well be meaningful to a botanist, but no interpretations will be attempted here.

Figure 12.1 shows a draftsman's diagram of the plot number (1–17) and the first three principal components. It is noticeable that the first component is closely related to the plot number. This reflects the fact that the plots are in order of the abundance in the plots of species with a high response to light and a low response to moisture, soil reaction, and nitrogen. Hence, the analysis has at least been able to detect this trend.

TABLE 12.1

Eigenvalues from a Principal Components Analysis of the Data in Table 9.7 Plots are objects of interest and the species counts are the variables.

Component	Eigenvalue	% of Total	Cumulative
1	8.79	35.20	35.20
2	5.59	22.30	57.50
3	2.96	11.80	69.30
4	1.93	7.70	77.00
5	1.58	6.30	83.40
6	1.13	4.50	87.90
7	0.99	4.00	91.90
8	0.55	2.20	94.00
9	0.40	1.60	95.60
10	0.35	1.40	97.00
11	0.20	0.80	97.80
12	0.18	0.70	98.50
13	0.13	0.50	99.00
14	0.12	0.50	99.50
15	0.07	0.30	99.80
16	0.05	0.20	100.00
Total	25.00	100.00	

TABLE 12.2

The First Three Principal Components for the Data in Table 9.7

Species	Z_1	Z_2	Z_3
Festuca ovina	−0.30	0.00	0.07
Anemone nemorosa	−0.25	−0.02	0.18
Stallaria holostea	−0.20	−0.20	0.19
Agrostis tenuis	0.17	−0.24	−0.01
Ranunculus ficaria	−0.11	0.32	0.07
Mercurialis perennis	−0.08	0.31	−0.02
Poa pratenis	−0.11	−0.32	0.11
Rumex acetosa	−0.01	−0.34	−0.23
Veronica chamaedrys	−0.15	−0.36	0.06

(*Continued*)

TABLE 12.2 (Continued)

Species	Z_1	Z_2	Z_3
Dactylis glomerata	−0.23	−0.15	−0.18
Fraxinus excelsior (juv.)	−0.25	0.11	−0.17
Saxifraga granulata	0.13	−0.24	−0.23
Deschampsia flexuosa	−0.05	−0.12	0.44
Luzula campestris	0.28	−0.09	0.00
Plantago lanceolata	0.27	−0.11	−0.26
Festuca rubra	−0.03	−0.23	−0.19
Hieracium pilosella	0.27	0.02	−0.05
Geum urbanum	−0.20	0.18	−0.29
Lathyrus montanus	−0.15	−0.26	0.19
Campanula persicifolia	−0.21	−0.18	−0.07
Viola riviniana	−0.23	−0.16	−0.11
Hepatica nobilis	−0.21	−0.03	−0.34
Achillea millefolium	0.29	−0.03	−0.10
Allium sp.	−0.18	0.12	−0.36
Trifolim repens	0.21	−0.11	−0.22

Note: The values shown are for the coefficients of the standardized species abundances, with means of zero and standard deviations of one.

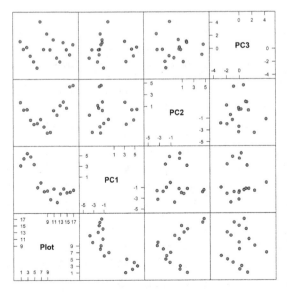

FIGURE 12.1
Draftsman's diagram for the ordination of 17 plots from Steneryd Nature Reserve.

12.2.2 Example 12.2: Burials in Bannadi

For a second example of principal components ordination, we will use the data shown in Table 9.8 concerning grave goods from a cemetery in Bannadi, northeast Thailand. The full table, supplied by Professor C.F.W. Higham, gives the presence or absence of 38 different types of articles in each of 47 burials, with additional information on whether the body was of an adult male, an adult female, or a child. In Exercise 2 of Chapter 9, cluster analysis was used to study the relationships between the burials. Now, ordination is considered with the same end in mind. For a principal components analysis, the burials are the objects of interest, and the 38 types of grave goods provide the 0–1 variables to be analyzed. These variables were standardized before use so that the analysis was based on their correlation matrix.

In a situation where only presence and absence data are available, it is common to find that a fairly large number of principal components are needed to account for most of the variation in the data. This is certainly the case here, with 11 components needed to account for 80% of the variance and 15 required to account for 90% of the variance. Obviously, there are far too many important principal components for a satisfactory ordination.

For this example, only the first four principal components will be considered, with the understanding that much of the variation in the original data is not accounted for. In fact, the four components correspond to eigenvalues of 5.29, 4.43, 3.65, and 3.34, while the total of all the eigenvalues is 38 (the number of types of articles). Thus, these components account for 13.9%, 11.6%, 9.6%, and 8.8%, respectively, of the total variance, and between them they account for 43.9% of the variance.

The coefficients of the standardized presence-absence variables are shown in Table 12.3 with the largest values (arbitrarily set at an absolute value greater than 0.2) underlined. To aid in interpretation, the signs of the coefficients have been reversed if necessary to ensure that the values of all the components are positive for burial B48, which has the largest number of items present. This is allowable because switching the signs of all the coefficients for a component does not change the percentage of variation explained by the component; the only effect is to possibly make the story-telling easier.

The large coefficients of Component 1 indicate the presence of articles type 18, 19, 20, 23, 25, 26, 32, 34, and 37, and the absence of articles type 3, 5, 6, 14, and 28. There is no grave with exactly this composition, but the component measures the extent to which each of the graves matches this model. The other components can also be interpreted in a similar way from the coefficients in Table 12.3.

Figure 12.2 shows a draftsman's plot of the total number of goods, the type of body, and the first four principal components. From studying this, it is possible to draw some conclusions about the nature of the graves. For example, it seems that male graves tend to have low values and female graves to have high values for principal component 1, possibly reflecting a difference

TABLE 12.3

Coefficients of Standardized Presence–Absence Data for the First Four Principal
Components of the Bannadi Data

Article	PC1	PC2	PC3	PC4
1	0.01	−0.02	0.01	0.00
2	−0.09	−0.04	−0.02	**0.52**
3	**−0.23**	**0.39**	−0.01	−0.03
4	−0.09	−0.04	−0.02	**0.52**
5	**−0.23**	**0.39**	−0.01	−0.03
6	**−0.23**	**0.39**	−0.01	−0.03
7	−0.02	−0.05	−0.02	−0.02
8	−0.03	−0.02	−0.02	−0.02
9	0.17	0.12	0.33	0.06
10	0.15	0.09	0.04	0.07
11	0.05	0.03	0.03	−0.02
12	0.12	0.04	0.11	0.05
13	−0.01	−0.05	−0.02	−0.09
14	**−0.23**	**0.39**	−0.01	−0.03
15	0.00	0.00	0.03	−0.01
16	0.17	0.12	**0.33**	0.06
17	−0.01	−0.05	−0.02	−0.09
18	**0.22**	0.15	**−0.38**	0.04
19	**0.22**	0.15	**−0.38**	0.04
20	**0.22**	0.15	**−0.38**	0.04
21	−0.09	−0.04	−0.02	**0.52**
22	0.00	−0.04	0.01	0.01
23	**0.27**	0.17	**−0.24**	0.08
24	0.03	0.03	0.05	−0.07
25	**0.26**	0.15	**0.28**	0.11
26	**0.26**	0.15	**0.28**	0.11
27	0.08	0.02	0.02	0.04
28	**−0.22**	0.19	−0.02	0.26
29	−0.17	0.17	0.01	−0.08
30	0.17	0.11	0.00	−0.05
31	0.08	0.03	0.18	−0.04
32	**0.27**	0.14	0.04	0.03
33	−0.02	0.00	0.06	−0.12
34	**0.23**	0.09	−0.10	0.03

(Continued)

TABLE 12.3   (Continued)

Article	PC1	PC2	PC3	PC4
35	0.04	0.01	0.17	0.14
36	−0.07	0.15	0.17	−0.08
37	**0.26**	0.11	0.07	0.02
38	0.12	**0.22**	0.05	−0.05

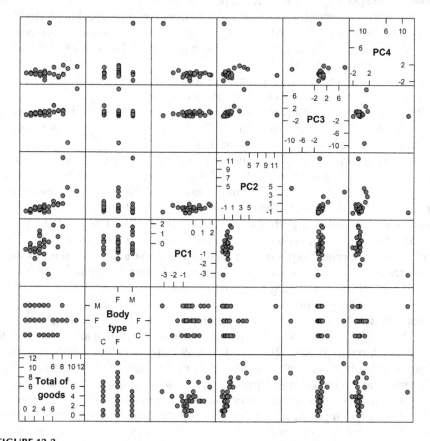

FIGURE 12.2
Draftsman's diagram for 47 Bannadi graves. The variables plotted are the total number of different types of goods, the type of remains (C = child, F = adult female, M = adult male) and the first four principal components.

in grave goods associated with sex. Also, grave B47 has an unusual composition in comparison with the other graves. However, the fact that four principal components are being considered makes a simple interpretation of the results difficult.

12.3 Principal Coordinates Analysis

Principal coordinates analysis (PCO) is an ordination method that can be thought of as a generalization of principal components analysis (PCA) and, at the same time, the method shares the goals of nonmetric multidimensional scaling (Chapter 11). At least as defined in this book, multidimensional scaling intends to represent in a map the configuration of distances between objects the best as possible. The representation in multidimensional scaling is sought to be optimal, and the optimality criterion is particularly judged in terms of the stress, described as the extent to which the positions of objects in a t-dimensional configuration fail to match the original distances or similarities after appropriate scaling.

PCO also addresses the problem of producing a map to make the distance between every pair of objects as nearly as possible equal to their distance, with the number of dimensions of that map taking the role of the number of axes constructed in PCA. We will briefly review here principal coordinates from the ideas learnt in Chapters 6 and 11.

In PCA, the original $p \times n$ data matrix (Table 6.2) is projected onto a space of fewer than p dimensions, with the scales of the axes governed by the eigenvectors of the covariance or correlation matrices. Principal components are formed by linear combinations of the original variables (when the covariance matrix is used) or perhaps those variables standardized to have mean zero and variance one (when the correlation matrix is considered). The first principal component explains as much of the variation in the data as possible, then the second axis (second component) is perpendicular to the first (a usual restriction on creating axes) such that it, in turn, explains as much as possible of the remaining variation in the data. And so on for the remaining components. One can correctly use somewhat more abstract language and say that PCA creates a new "coordinate system" of dimension t, with $t < p$, for the original n objects. This coordinate system "lives" in Euclidean space. Thus, a plot of objects for the first two principal components lies on a two-dimensional Euclidean space, providing a map of the objects on the plane.

Like nonmetric multidimensional scaling (Chapter 11), PCO constructs a configuration of n points in the Euclidean space using information about the distances between the n objects. PCO can start with *any* $n \times n$ matrix $\mathbf{D}$ of distances between n objects, and endeavors to find t ordination axes in such a way as to make the distance between every pair of objects in t dimensions as nearly as possible equal to their original distance or similarity given by $\mathbf{D}$. The number of axes of the new coordinate system in which the objects will be projected depends only on the number of objects, n.

The solution to PCO described was given by Schoenberg (1935) and Young and Householder (1938), and later refined by Torgerson (1958) and Gower (1966). Assume that it is intended to choose a configuration (coordinates) in the t-dimensional space. The PCO algorithm commences with an $n \times n$

distance matrix **D**, containing the distances d_{ik} between objects i and k, being transformed into a new $n \times n$ matrix **A**, such that

$$a_{ik} = -d_{ik}^2 / 2 \qquad (12.1)$$

The algorithm continues with a transformation of **A**, known as *double-centering*, to position the origin of the new coordinate system. The element occupying the ith row and kth column of the $n \times n$ double-centered matrix **B** is given by

$$b_{ik} = a_{ik} - \bar{a}_{i\bullet} - \bar{a}_{\bullet k} + \bar{a}_{\bullet\bullet}, \qquad (12.2)$$

where $\bar{a}_i$ is the mean of the ith row of **A**, $\bar{a}_k$ is the mean of the kth column of **A**, and $\bar{a}..$ is the mean of all the elements in **A**. The double-centered matrix **B** will have zero row and column means and is, therefore, more suitable for the analysis. It can be shown (see Mardia et al., 1979) that the principal coordinates are determined by the first t eigenvectors of **B**. If t is low (say, 2, 3 or 4) and the corresponding t eigenvalues of **B** are relatively large and positive, and the other are positive but small, or negative, then it is expected that the distances between objects of this t-dimensional configuration will approximate **D**.

Principal coordinates analysis can also start with a matrix **S** containing similarities between objects i and k, s_{ik}, for $i, k = 1, 2, \ldots, n$, using any of the many available similarity indices. The algorithm described here requires s_{ik} be transformed and summarized in a suitable distance matrix **D**. The *standard transformation* of s_{ik} specifies that the distance between objects i and k must be:

$$d_{ik} = \sqrt{s_{ii} + s_{kk} - 2s_{ik}}. \qquad (12.3)$$

By letting $a_{ik} = s_{ik}$, it is easily seen that d_{ik}, as defined by (12.3), enjoys the property expressed in Equation (12.1).

The use of similarity matrices in principal coordinates analysis justifies the connection between this method and principal components analysis. Assume that the $n \times p$ data matrix **X** (like Table 6.2) is known, and its columns are then centered to produce a new data matrix $\mathbf{X}_C$ (the mean of each variable is subtracted from each observed value for that variable, forcing the variables to have zero means). If principal coordinates analysis is applied to the symmetric matrix $\mathbf{S} = \mathbf{X}_C' \mathbf{X}_C$, the results will produce essentially the same ordination as principal components analysis on the data in **X** by noticing that $(n-1)\mathbf{C} = \mathbf{S} = \mathbf{X}_C' \mathbf{X}_C$, where **C** is the covariance matrix of the X variables. The elements of $\mathbf{S} = \mathbf{X}_C' \mathbf{X}_C$ can be thought as measures of similarities between the n objects being considered. Indeed, if s_{ik} is the element in the ith row and kth column of $\mathbf{X}'_C\mathbf{X}_C$, s_{ik} is a measure of the similarity between objects i and k, because increasing s_{ik} in (12.3) means that the Euclidean distance d_{ik} between the objects is decreased. Further, also from (12.3) it is seen that s_{ik} takes the maximum value of $(s_{ii} + s_{kk})/2$ when $d_{ik} = 0$, which occurs when the objects i and k have identical values for the variables X_1 to X_p. Therefore, using the centered data makes the result of principal coordinates interpretable as being equivalent to a principal component analysis applied to $(n-1)$ times the covariance matrix of the X variables. The only difference will be in terms of the scaling given to the components. In principal components analysis, it is usual to scale the lth component to have the variance λ_l, but with a principal coordinates analysis, the component would usually be scaled to have a variance of $(n-1)\lambda_l$. This difference is immaterial because it is only the relative values of objects on ordination axes that are important.

When the PCO algorithm described is based on the matrix of Euclidean distances between the ith and the kth objects of the raw data $\mathbf{X}$

$$d_{ik} = \sqrt{\sum_{j=1}^{p}(x_{ij} - x_{kj})},$$

it can be verified that the result will be the same as a principal component analysis based on the covariance matrix for the same data, up to the scaling factor $(n-1)$. Moreover, the double-centered matrix (Equation 12.2) is expressible as $\mathbf{B} = \mathbf{X}_C\mathbf{X}'_C$. All these properties emphasize the strong links between PCO and PCA.

There are two complications that can arise in principal coordinates analysis. First, the double-centered matrix $\mathbf{B}$ always has one or more null eigenvalues. This is a consequence of the centering. The second complication is that some of the eigenvalues of the distance matrix may be negative. This is disturbing because the corresponding principal components appear to have negative variances! However, the truth is just that the distance or similarity matrix could not have been obtained by means of the expressions (12.1) or (12.3) or, equivalently, that it is impossible to arrange n points in a space of any number of dimensions with the pairwise distances exactly the same as the values in $\mathbf{B}$. With ordination, only the components associated with the largest eigenvalues are usually used, so that a few small negative eigenvalues can be regarded as being unimportant. Large negative eigenvalues suggest that the distance or similarity matrix being used is not suitable for ordination.

Computer programs for principal coordinates analysis usually offer the option of starting with either a distance matrix or a similarity matrix. If a similarity matrix is used, then it can be converted to a distance matrix by transforming the similarity s_{ik} to the distance measure (12.3) or any other valid transformation.

12.3.1 Example 12.3: Plant Species in the Steneryd Nature Reserve (Revisited)

As an example of the use of principal coordinates analysis, the data considered in Example 12.1 on species abundances on 17 plots in Steneryd Nature Reserve were reanalyzed using Manhattan distances between plots. That is, the distance between plots i and k was measured by $d_{ik} = \Sigma \left| x_{ij} - x_{kj} \right|$, where the summation is for j over the 25 species and x_{ij} denotes the abundance of species j on plot i, as given in Table 9.7. These distances were transformed as $a_{ik} = -d_{ik}^2 / 2$, placed into a 17×17 matrix $\mathbf{A}$, and then double-centered before eigenvalues and eigenvectors were calculated.

The first two eigenvalues of the double-centered matrix were found to be 100,081 and 55,711, which account for 47.0% and 26.2% of the sum of the eigenvalues, respectively. On the face of it, the first two axes, therefore, give a good ordination, with 73.2% of the variation accounted for. The third eigenvalue is much smaller at 19,900 and only accounts for 9.3% of the total.

Figure 12.3 shows a draftsman's diagram of the plot number and the first two principal coordinates, with the sign of the second coordinate being reversed to allow the comparison of results with the ordination of the same data using principal components (Example 12.1). Both PCO axes show a relationship with the plot number, which, as noted in Example 12.1, is itself related to the response of the different species to environmental variables. A comparison of this diagram with the six graphs in the bottom left-hand corner of Figure 12.1 shows that the first two axes from the principal coordinates analysis are really very similar to the first two principal components apart from a difference in scaling.

12.3.2 Example 12.4: Burials in Bannadi (Revisited)

As an example of a principal coordinates analysis on presence and absence data, consider again the data in Table 9.8 on grave goods in the Bannadi cemetery in northeast Thailand. The analysis was carried out using R (R Core Team, 2024) and started with the matrix of unstandardized Euclidean distances between the 47 burials so that the distance from grave i to grave k was taken to be $d_{ik} = \sqrt{\Sigma (x_{ij} - x_{kj})^2}$, where the summation is for j from 1 to 38, and x_{ij} is 1 if the jth type of article is present in the ith burial or is otherwise 0. The transformed matrix $\mathbf{A}$ was then obtained as described in Example 12.3 and double-centered before eigenvalues and eigenvectors were obtained.

The principal coordinates analysis carried out in this manner gives the same result as a principal components analysis using unstandardized values for the X variables (i.e., carrying out a principal components analysis using the sample covariance matrix instead of the sample correlation matrix). The only difference in the results is in the scalings that are usually given to the ordination variables by principal components analysis and principal

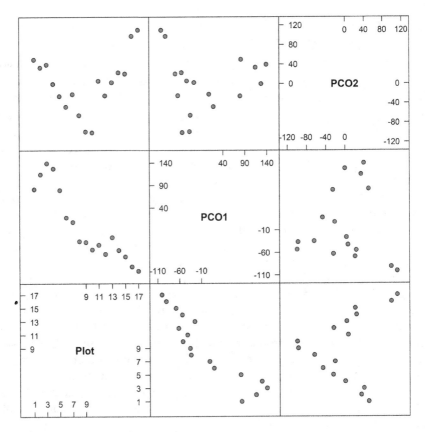

FIGURE 12.3
Draftsman's diagram for the ordination of 17 plots from Steneryd Nature Reserve based on a principal coordinates analysis on Manhattan distances between plots. The three variables are the plot number and the first two principal coordinates (PCO1 and PCO2). The sign of PCO2 were reversed from those generated by the computer output to permit the comparison of results to the ordination based on principal components (Example 12.1).

coordinates analysis. The first four eigenvalues of the double-centered matrix were 24.9, 19.2, 10.0, and 8.8, corresponding to 21.5%, 16.6%, 8.7%, and 7.6%, respectively, of the sum of all the eigenvalues. These four principal coordinates account for a mere 54.4% of the total variation in the data, but this is better than the 43.9% accounted for by the first four principal components obtained from the standardized data (Example 12.2).

Figure 12.4 shows a draftsman's diagram for the total number of goods in the burials, the type of body (child, adult female, or adult male), and the first four components. The signs of the second and third principal coordinates were switched from those shown on the computer output so as to make them have positive values for burial B48, which contained the largest number of different types of grave goods. It can be seen from the diagram that

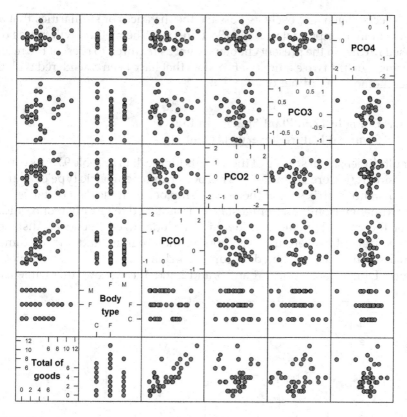

FIGURE 12.4
Draftsman's diagram for the 47 Bannadi graves. The six variables are the total number of different types of goods in a burial, an indicator of the type of remains (C = child, F = adult female, M = adult male), and the first four components from a principal coordinates analysis (PCO1 to PCO4).

the first coordinate represents total abundance quite closely, but the other principal coordinates are not related to this variable. Apart from this, the only obvious thing to notice is that one of the burials had a very low value for the fourth component. This is burial B47, which contained eight different types of article, of which four types were not seen in any other burial.

12.4 Multidimensional Scaling

Multidimensional scaling has been discussed already in Chapter 11, where it was defined as an iterative process for finding coordinates for objects on axes, in a specified number of dimensions, such that the distances between the objects match as closely as possible the distances or similarities that are provided in an input data matrix (Section 11.2). The method will not be discussed further in the present chapter except as required to present the results of using it on the two example sets of data that have been considered with the other methods of ordination.

12.4.1 Example 12.5: Plant Species in the Steneryd Nature Reserve (Again)

A multidimensional scaling of the 17 plots for the data in Table 9.7 was carried out using R (R Core Team, 2024). This performs a nonmetric type of analysis on a distance matrix so that the relationship between the data distances and the ordination (configuration) distances is assumed to be only monotonic.

For the example being considered, unstandardized Euclidean distances between the plots were used as input to the program, and a three-dimensional solution was assumed. Figure 12.5 shows the plot numbers plotted against Dimensions 1 and 2, 1 and 3, and 2 and 3. The shows that Dimension

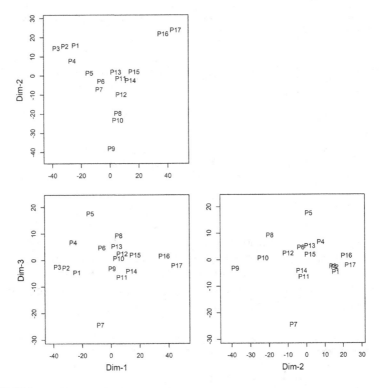

FIGURE 12.5
Results from the ordination of 17 plots from Steneryd Nature Reserve with nonmetric multidimensional scaling based on Euclidean distances between the plots using unstandardized data, assuming a three-dimensional solution (Dim-1 to Dim-3).

1 is strongly related to the plot number, while Dimension 2 indicates that the central plots differ to some extent from the plots with high and low numbers.

12.4.2 Example 12.6: Burials in Bannadi (Again)

The same analysis as used in the last example was also applied to the data on burials at Bannadi shown in Table 9.8. Unstandardized Euclidean distances between the 47 burials were calculated using the 0–1 data in the table as values for 38 variables, and these distances provided the data for R. Figure 12.6 shows the burial numbers with the body types juxtaposed against Dimensions 1 and 2, 1 and 3, and 2 and 3. This shows the burials with the highest number of goods around the outside of the plots, burials that mostly contain adult female and male bodies. Only two of these burials, B38 and B19, contain child bodies; they are also characterized for having the two highest number of goods among child graves. The centers of the plots contain the burials with few goods.

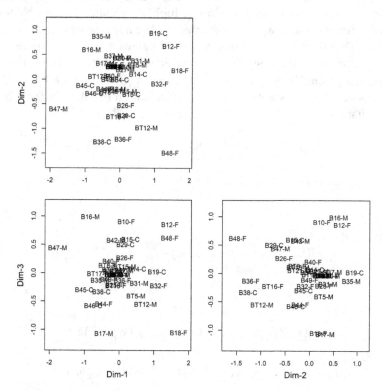

FIGURE 12.6
Plot of the values for the three axes from non-metric multidimensional scaling using unstandardized Euclidean distances between the Bannadi graves (Dim-1 to Dim-3). Each point is labeled by the burial number and the type of body type found in the burial: child (C), adult female (F), adult male (M).

12.5 Correspondence Analysis

Correspondence analysis as a method of ordination originated in the work of Hirschfeld (1935), Fisher (1940), and a school of French statisticians (Benzecri, 1992). It is a popular method of ordination for plant ecologists and is used in other areas as well.

The method will be explained here in the context of the ordination of p sites on the basis of the abundance of n species, although it can be used equally well on data that can be presented as a two-way table of measures of abundance with the rows corresponding to one type of classification and the columns to a second type of classification.

With sites and species, the situation is as shown in Table 12.4. Here, there are a set of species values $a_1, a_2, ..., a_n$ associated with the rows of the table, and a set of site values $b_1, b_2, ..., b_p$ associated with the columns of the table. One interpretation of correspondence analysis is that is concerned with choosing species and site values so that they are as highly correlated as possible for the bivariate distribution that is represented by the abundances in the body of the table. That is to say, the site and species values are chosen to maximize their correlation for the distribution in which the number of times that species i occurs at site j is proportional to the observed abundance x_{ij}.

It turns out that the solution to this maximization problem is given by the set of equations

$$a_1 = \left[(x_{11} / R_1)b_1 + (x_{12} / R_1)b_2 + ... + (x_{1p} / R_1)b_p \right] / r$$

$$a_2 = \left[(x_{21} / R_2)b_1 + (x_{22} / R_2)b_2 + ... + (x_{2p} / R_2)b_p \right] / r$$

$$\vdots$$

$$a_n = \left[(x_{n1} / R_n)b_1 + (x_{n2} / R_n)b_2 + ... + (x_{np} / R_n)b_p \right] / r$$

and

$$b_1 = \left[(x_{11} / C_1)a_1 + (x_{21} / C_1)a_2 + ... + (x_{n1} / C_1)a_n \right] / r$$

$$b_2 = \left[(x_{12} / C_2)a_1 + (x_{22} / C_2)a_2 + ... + (x_{n2} / C_2)a_n \right] / r$$

$$\vdots$$

$$b_p = \left[(x_{1p} / C_p)a_1 + (x_{2p} / C_p)a_2 + ... + (x_{np} / C_p)a_n \right] / r$$

where

R_i denotes the total abundance of species i,

R_j denotes the total abundance at site j, and

r is the maximum correlation being sought.

Thus, the ith species value a_i is a weighted average of the site values, with site j having a weight that is proportional to x_{ij}/R_i, and the jth site value b_j is a weighted average of the species values, with species i having a weight that is proportional to x_{ji}/C_j.

The name *reciprocal averaging* is sometimes used to describe the equations just stated because the species values are (weighted) averages of the site values, and the site values are (weighted) averages of the species values. These equations are themselves often used as the starting point for justifying correspondence analysis as a means of producing species values as a function of site values, and vice versa. It turns out that the equations can be solved iteratively after they have been modified to remove the trivial solution with $a_i = 1$ for all i, $b_j = 1$ for all j, and $r = 1$. However, it is more instructive to write the equations in matrix form to solve them because this shows that there may be several possible solutions to the equations and that these can be found from an eigenvalue analysis.

In matrix form, these equations just presented become

$$a = R^{-1}Xb / r \qquad (12.4)$$

and

$$b = C^{-1}X'a / r \qquad (12.5)$$

where

$$a' = (a_1, a_2, ..., a_n),$$

$$b' = (b_1, b_2, ..., b_p),$$

R is an n by n diagonal matrix, with R_i in the ith row and ith column,

C is a p by p diagonal matrix with C_j in the jth row and jth column, and

X is an n by p matrix with x_{ij} in the ith row and jth column.

If Equation 12.5 is substituted into Equation 12.4, then, after some matrix algebra, it is found that

$$(R^{-1/2}XC^{-1/2})(R^{-1/2}XC^{-1/2})'(R^{-1/2}a) = r^2(R^{-1/2}a) \qquad (12.6)$$

where

$\mathbf{R}^{1/2}$ is a diagonal matrix with $\sqrt{R_i}$ in the ith row and ith column and

$\mathbf{C}^{1/2}$ is a diagonal matrix with $\sqrt{C_j}$ in the jth row and jth column.

This shows that the solutions to the problem of maximizing the correlation are given by the eigenvalues of the n by n matrix:

$$(\mathbf{R}^{-1/2}\mathbf{X}\mathbf{C}^{-1/2})(\mathbf{R}^{-1/2}\mathbf{X}\mathbf{C}^{-1/2})'.$$

For any eigenvalue λ_k, the correlation between the species and site scores will be $r_k = \sqrt{\lambda_k}$, and the eigenvector for this correlation will be

$$\mathbf{R}^{1/2}\mathbf{a}_k = (\sqrt{R_1}\,a_{1k}, \sqrt{R_2}\,a_{2k}, ..., \sqrt{R_n}\,a_{nk})'$$

where a_{ik} are the species values. The corresponding site values can be obtained from Equation 12.5 as

$$\mathbf{b}_k = \mathbf{C}^{-1}\mathbf{X}'\mathbf{a}_k / r_k.$$

The largest eigenvalue will always be $r^2 = 1$, giving the trivial solution $a_i = 1$ for all i and $b_j = 1$ for all j. The remaining eigenvalues will be positive or zero and reflect different possible dimensions for representing the relationships between species and sites. These dimensions can be shown to be orthogonal in the sense that the species and site values for one dimension will be uncorrelated with the species and site values in other dimensions for the data distribution of abundances x_{ij}.

Ordination by correspondence analysis involves using the species and site values for the first few largest eigenvalues that are less than 1 because these are the solutions for which the correlations between species values and site values are strongest. It is common to plot both the species and the sites on the same axes because, as noted earlier in this section, the species values are an average of the site values, and vice versa. In other words, correspondence analysis gives an ordination of both species and sites at the same time.

It is apparent from Equation 12.6 that correspondence analysis cannot be used on data that include a zero-row sum because then the diagonal matrix $\mathbf{R}^{-1/2}$ will have an infinite element. By a similar argument, zero column sums are not allowed either. This means that the method cannot be used on the burial data in Table 9.8 because some graves did not contain any articles. However, correspondence analysis can be used with presence and absence data when this problem is not present.

12.5.1 Example 12.7: Plant Species in the Steneryd
Nature Reserve (Yet Again)

Correspondence analysis was applied to the data for species abundances in the Steneryd Nature Reserve (Table 9.7). Only the first two dimensions were considered for the ordination plot. Figure 12.7 shows a graph of the species, with abbreviated names, the site numbers, and Dimensions 1 and 2. The ordination of the plots is quite clear, with an almost perfect sequence from Site 17 on the left (S17) to Site 1 on the right, moving around the very distinct arch. The species are interspersed among the sites along the same arch, from *Mercurialis perennis* (*Mer per*) on the left to *Hieracium pilosella* (*Hie pil*) on the right. A comparison of the figure with Table 9.7 shows that this makes a good deal of sense. For example, *M. perennis* is only abundant on the highest-numbered sites, and *H. pilosella* is only abundant on the lowest-numbered sites.

The arch or horseshoe that appears in the ordination for this example is a common feature in the results of correspondence analysis, which is also sometimes apparent with other methods of ordination.

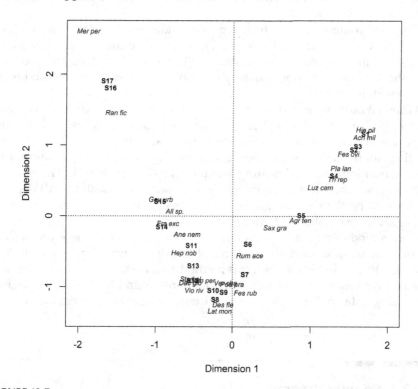

FIGURE 12.7
Plots of species and site numbers against the first two axes (Dimensions 1 and 2) found by applying correspondence analysis to the data from Steneryd Nature Reserve. The species names (in italics) have obvious abbreviations and the sites (in boldface) are numbered from S1 to S17.

There is sometimes concern that this effect will obscure the nature of the ordination axes, and therefore, some attention has been devoted to the development of ways to modify analyses to remove the effect, which is considered to be an artefact of the ordination method. With correspondence analysis, a method of detrending is usually used, and the resulting ordination method is then called *detrended correspondence analysis* (Hill and Gauch, 1980). Adjustments for other methods of ordination exist as well but seem to receive little use.

12.6 Comparison of Ordination Methods

Four methods of ordination have been reviewed in this chapter. Unfortunately, there is no satisfactory way to state when each should be used because of the wide variety of different circumstances for which ordination is used. Therefore, all we do here is make some final comments on each of the methods in terms of its utility.

Principal components analysis requires values for p variables for each of the objects being studied. Therefore, PCA cannot be used when only a distance or similarity matrix is available. When variable values are available and the variables are approximately normally distributed, this method is an obvious choice.

When an ordination is required starting with a matrix of distance or similarities between the objects being studied, it is possible to use either principal coordinates analysis or multidimensional scaling. Multidimensional scaling can be metric or nonmetric, and principal coordinates analysis and metric multidimensional scaling should give similar results. The relative advantages of metric and nonmetric multidimensional scaling will depend very much on the circumstances, but, in general, nonmetric scaling can be expected to give a slightly better fit to the data matrix.

Correspondence analysis was developed for situations where the objects of interest are described by measures of the abundance of different characteristics. When this is the case, this method appears to give ordinations that are relatively easy to interpret. It has certainly found favor with ecologists analyzing data on the abundance of different species at different locations.

12.7 Computer Programs

The analyses described in this chapter can be carried out using many standard statistical packages, and there are also some specialized packages developed for specific areas of application such as for analyzing plant distribution data. Also, the appendix to this chapter explains how the various ordination analyses described in the chapter can be carried out using packages in R.

12.8 Further Reading

The books by Legendre and Legendre (2012), Greenacre and Primicerio (2013), Gotelli and Ellison (2018), and Thioulouse et al. (2018) cover ordination methods for ecologists in more detail. Suggestions for further reading related to principal components analysis and multidimensional scaling are provided in Chapters 6 and 11, and it is unnecessary to repeat these here. For further discussions and more examples of principal coordinates analysis and correspondence analysis, most of them in the context of plant ecology, see the books by Digby and Kempton (1987), Ludwig and Reynolds (1988), Jongman et al. (1995), and Šmilauer and Lepš (2014).

For correspondence analysis, the most recommended reference is by Greenacre (2017). In addition, there is a short book on correspondence analysis by Clausen (1998) and a very comprehensive book on the same topic by Benzecri (1992). In the spirit of the French school, the book by Husson et al. (2017) covers methods of multivariate exploratory data analysis, essentially principal components, correspondence analysis and its extension, multiple correspondence analysis.

One important technique not covered in this chapter is canonical ordination, in which the ordination axes are chosen to represent a set of explanatory variables as much as possible. For example, there might be interest in seeing how the distribution of plant species over a number of sites is related to the temperature and soil characteristics at those sites. Discriminant function analysis is one special case of this type of analysis, but a number of other analyses are also possible. See Legendre and Legendre (2012) for more details.

Exercise

Table 6.8 shows the values for six measurements taken on each of 25 prehistoric goblets excavated in Thailand. The nature of the measurements is shown in Figure 6.6. Use the various methods discussed in this chapter to produce ordinations of the goblets and see which method appears to produce the most useful result.

References

Benzecri, P.J. (1992). *Correspondence Analysis Handbook*. New York: Marcel Dekker.

Clausen, S.E. (1998). *Applied Correspondence Analysis*. Thousand Oaks, CA: SAGE.

Digby, P.G.N., and Kempton, R.A. (1987). *Multivariate Analysis of Ecological Communities*. London: Chapman and Hall.

Fisher, R.A. (1940). The precision of discriminant functions. *Annals of Eugenics* 10: 422–429.

Gotelli, N.J., and Ellison, A.M. (2018). *A Primer of Ecological Statistics*. 2nd Edn. Sunderland, MA: Sinauer Associates Inc.

Gower, J.C. (1966). Some distance properties of latent root and vector methods in multivariate analysis. *Biometrika* 53: 315–328.

Greenacre, M. (2017). *Correspondence Analysis in Practice*. 3rd Edn. Boca Raton, FL: CRC Press.

Greenacre, M., and Primicerio, R. (2013). *Multivariate Analysis of Ecological Data*. Bilbao: Fundación BBVA. www.fbbva.es/wp-content/uploads/2017/05/dat/ DE_2013_multivariate.pdf.

Hill, M.O., and Gauch, H.G. (1980). Detrended correspondence analysis, an improved ordination technique. *Vegetatio* 42: 47–58.

Hirschfeld, H.O. (1935). A connection between correlation and contingency. *Proceedings of the Cambridge Philosophical Society* 31: 520–524.

Jongman, R.H.G., ter Braak, C.J.F., and van Tongeren, O.F.F. (1995). *Data Analysis in Community and Landscape Ecology*. Cambridge: Cambridge University Press.

Legendre, P., and Legendre, L. (2012). *Numerical Ecology*. 3rd Edn. Amsterdam: Elsevier.

Ludwig, J.A., and Reynolds, J.F. (1988). *Statistical Ecology*. New York: Wiley.

Mardia, K.V., Kent, J.T., and Bibby, J.M. (1979). *Multivariate Analysis*. London: Academic Press.

R Core Team. (2024). *R: A Language and Environment for Statistical Computing*. Vienna: R Foundation for Statistical Computing. www.r-project.org/.

Schoenberg, I.J. (1935). Remarks to Maurice Frechet's article "Sur la definition axiomatique d'une classe d'espace distanciés vectoriellement applicable sur l'espace de Hilbert." *Annals of Mathematics* 36: 724–732.

Šmilauer, P., and Lepš, J. (2014). *Multivariate Analysis of Ecological Data Using Canoco 5*. 2nd Edn. Cambridge: Cambridge University Press.

Torgerson, W.S. (1958). *Theory and Methods of Scaling*. New York: Wiley.

Thioulouse, J., Dray, S., Dufour, A., Siberchicot, A., Jombart, T., and Pavoine, S. (2018). *Multivariate Analysis of Ecological Data with ade4*. New York: Springer.

Young, G., and Householder, A.S. (1938). Discussion of a set of points in terms of their mutual distances. *Psychometrika* 3: 19–22.

Appendix: Ordination Methods in R

Fully commented R scripts, reproducing the results for each example presented in this chapter, are available at the book's website. Our strategy in writing all those scripts was to use the simplest R commands for that purpose; at the end of this appendix, we suggest additional R packages offering a wider range of computational procedures for ordination analysis.

A.1 Principal Components and Nonmetric Multidimensional Scaling

It was indicated in this chapter that principal components analysis (Section 12.2) and nonmetric multidimensional scaling (Section 12.4) belong to the collection of eligible methods that a data analyst may apply for producing the ordination of multivariate data. Consequently, we refer the reader to the appendices in Chapters 6 and 11 to select the R commands needed to generate the reduced set of variables from principal components analysis or nonmetric multidimensional scaling.

A.2 Principal Coordinates Analysis

Principal coordinates analysis can be run in R with cmdscale, a function whose name stands for classical multidimensional scaling. The main argument for this function is a distance matrix computed by the dist or any other equivalent function. As an example, given a data frame (say, MV.DATA) containing the variables of interest, and assuming that the Manhattan distance has been chosen to measure the distances among objects in MV.DATA, the distance matrix can be obtained as

```
Dist.Manh <- dist(MV.DATA, method = "manhattan")
```

Now, suppose you desire the maximum dimension of the reduced space to be $k = 3$ and that eigenvalues should be returned. The cmdscale command that has to be executed is then

```
PCO.object <- cmdscale(Dist.Manh, k = 3, eig = TRUE)
```

The user does not need to worry about transforming the distance matrix into a double-centered matrix as described in Section 12.3. The cmdscale command automatically computes that for you.

In the vegan package, there is the function capscale, which is a constrained version of principal coordinates analysis but can be used for classic PCO. It can place both the objects and the variables in a biplot-like figure.

A.3 Correspondence Analysis

The R command for correspondence analysis (CA), as described in the present chapter, is ca from the package ca (Nenadic and Greenacre, 2007). This command performs simple CA, a term given by the French school to mean that the data of interest in this case are usually a two-way contingency table. The terms multiple and joint correspondence analysis are used to denote extensions of simple CA to more than two categorical variables. See the paper by Nenadic and Greenacre (2007) for further details.

Once a simple CA is carried out for a data matrix mat.dat, for example, by

```
object.ca < - ca(mat.dat)
```

the user can get the complete set of eigenvalues by simply typing

```
summary(object.ca)
```

In addition, the correlations between row and column scores (or singular values) can be invoked with

```
object.ca$sv
```

A joint plot of the CA scores (see Figure 12.7) is then produced with the plot function

```
plot(object.ca)
```

A versatile alternative to the ca package is FactoMineR (Lê et al., 2008; Husson et al., 2017). This package implements not only commands for correspondence analysis (the CA function) but also includes a larger array of exploratory multivariate methods, like principal components, multiple factor analysis and multiple correspondence analysis.

A.4 Other R Packages for Ordination

Among the specialized R procedures for ordination analyses, the two particular packages labdsv (Roberts, 2023) and vegan (Oksanen et al., 2024) are very good. Both packages offer a diversity of ordination analyses created primarily to suit the computational needs of community ecologists, including principal components analysis, principal coordinates analysis, and multidimensional scaling. The scope of functions is larger in vegan than in labdsv. Thus, vegan allows the selection of two popular correspondence analysis algorithms, single CA and canonical CA, this latter known also as constrained CA, through the single command cca, while detrended correspondence analysis is invoked with the function decorana. Several comprehensive books and tutorials about the R functions for ordination analysis using vegan are available. We recommend the book by Borcard et al. (2018) and the tutorial (vignette) given by Oksanen (2024), the creator of the vegan package.

Ecologists and ecological statisticians have contributed to other R-packages for ordination analysis, all of them contained in a larger list of R packages for ecology available at https://drmattg.github.io/REcol-verse/. From this list, we recommend ade4 (Dray and Dufour, 2007; Thioulouse et al., 2018) and BiodiversityR (Kindt and Coe, 2005). These packages include all the ordination methods covered in this book and excel from others for they provide good quality tutorials.

A.5 Draftsman's Plots in Ordination

A general strategy suggested in the present chapter is to produce draftsman's diagrams as visual aids for drawing conclusions from any ordination analysis. These diagrams show scatterplots for a selected subset of variables derived by the multivariate analysis (the first few principal components in principal components analysis, the first few dimensions in PCO or nonmetric multidimensional scaling, etc.), and they may include a particular set of variables from the original data set, helpful in the interpretation of the ordination results. As seen in Chapter 3, whenever a draftsman's diagram is desired, the R functions pairs and scatterplotMatrix from the car package and splom from the lattice package are suitable commands that can be added to the R scripts in ordination analysis.

References

Borcard, D., Gillett, F., and Legendre, P. (2018). *Numerical Ecology in R*. 2nd Edn. New York: Springer.

Dray, S., and Dufour, A. (2007). The ade4 package: implementing the duality diagram for ecologists. *Journal of Statistical Software* 22(4): 1–20. https://doi.org/10.18637/jss.v022.i04.

Husson, F., Lê, S., and Pagès, J. (2017). *Exploratory Multivariate Analysis by Example Using R*. 2nd Edn. Boca Raton, FL: CRC Press.

Jongman, R.H.G., ter Braak, C.J.F., and van Tongeren, O.F.F. (1995). *Data Analysis in Community and Landscape Ecology*. Cambridge: Cambridge University Press.

Kindt, R., and Coe, R. (2005). *Tree Diversity Analysis. A Manual and Software for Common Statistical Methods for Ecological and Biodiversity Studies*. World Agroforestry Centre (ICRAF). www.worldagroforestry.org/output/tree-diversity-analysis.

Lê, S., Josse, J., and Husson, F. (2008). FactoMineR: an R package for multivariate analysis. *Journal of Statistical Software* 25(1): 1–18. https://doi.org/10.18637/jss.v025.i01.

Nenadic, O., and Greenacre, M. (2007). Correspondence analysis in R, with two- and three-dimensional graphics: the ca package. *Journal of Statistical Software* 20(3): 1–13.

Oksanen, J. (2024). *Vegan: An Introduction to Ordination*. https://cran.r-project.org/web/packages/vegan/vignettes/intro-vegan.pdf.

Oksanen, J., Simpson, G., Blanchet, F., Kindt, R., Legendre, P., Minchin, P., O'Hara, R., Solymos, P., Stevens, M., Szoecs, E., Wagner, H., Barbour, M., Bedward, M., Bolker, B., Borcard, D., Carvalho, G., Chirico, M., De Caceres, M., Durand, S., Evangelista, H., FitzJohn, R., Friendly, M., Furneaux, B., Hannigan, G., Hill, M., Lahti, L., McGlinn, D., Ouellette, M., Ribeiro Cunha, E., Smith, T., Stier, A., Ter Braak, C., and Weedon, J. (2024). *Vegan: Community Ecology Package*. R package version 2.6-6.1. https://CRAN.R-project.org/package=vegan.

Roberts, D.W. (2023). *labdsv: Ordination and Multivariate Analysis for Ecology*. R package version 2.1-0. https://CRAN.R-project.org/package=labdsv.

Thioulouse, J., Dray, S. Dufour, A., Siberchicot, A., Jombart, J., and Pavoine S. (2018). *Multivariate Analysis of Ecological Data with ade4*. New York: Springer.

13

Epilogue

13.1 The Next Step

In each edition of this book, we have purposely limited the content. Our aims will have been achieved if someone who has read the previous chapters carefully has a fair idea of what can and what cannot be achieved by the most widely used multivariate statistical methods. Our hope is that the book will help many people take the first step in "a journey of a thousand miles."

Those who have taken this first step possibly are interested in taking responsibilities as data analysts in their areas of expertise and seek to be competent in multivariate data analysis and multivariate data visualization. As with other areas of applied statistics, that competence requires practice. Thus, the way forward is to gain experience of multivariate methods by analyzing different sets of data and seeing what results are obtained.

There have been recent multivariate analysis developments in the closely related field of data mining, which is concerned with extracting information from very large data sets. Principal components, multidimensional scaling, and cluster analysis are common data mining exploration techniques. The development of data mining methods has rocketed because of the interconnection between the explosive availability of large data sets in all areas of human knowledge and technology, and the accessibility to high-capacity data processing software. One of the purposes of choosing the popular R system (R Core Team, 2024) in this book is to provide a gentle experience in the use of software for multivariate analysis with the capacity to cope with "big" data, one feature the R system has. The lesson to learn here is that the basic goals of each multivariate method will remain the same even for large data sets. The topic of data mining and the ampler discipline of data analytics should be investigated by anyone dealing with large multivariate data sets. More details will be found in the books by Hand et al. (2001) and Han et al. (2023).

DOI: 10.1201/9781003453482-13

13.2 Some General Reminders

There are a few general points to keep in mind when developing statistical analysis skills, including multivariate analyses. First, there are often alternative ways of approaching the analysis of a particular set of data, none of which is necessarily the best. Indeed, several types of analysis may well be carried out to investigate different aspects of the same data. For example, the body measurements of female sparrows given in Table 1.1 can be analyzed by principal components analysis or factor analysis to investigate the dimensions behind body size variation, by discriminant analysis to contrast survivors and nonsurvivors, by cluster analysis or multidimensional scaling to see how the birds group together, and so on.

Second, use common sense. Before embarking on an analysis, consider whether it can answer the questions of interest. Many statistical analyses are carried out because the data have a certain form, irrespective of what light the analyses can throw on a question. At some time or another, most users of statistics find themselves staring a large amount of computer output with the realization that it tells them nothing that they really want to know.

Third, multivariate analysis does not always produce a neat answer. There is an obvious bias in statistical textbooks and articles toward examples in which results are straightforward and conclusions are clear. In real life, this does not happen quite as often. Do not be surprised if multivariate analyses fail to give satisfactory results on the data that you are really interested in! There might well be a message in the data, but deciphering the message might require creative analyses. For example, it may be that variation in a multivariate set of data can be completely described by two or three underlying factors. However, these may not show up in a principal components analysis or a factor analysis because the relationship between the observed variables and the factors is not a simple linear one.

Finally, an analysis might be dominated by one or two rather extreme observations. These outliers can sometimes be found by data exploration techniques: frequency tables for the distributions of individual variables or various graphs. In some cases, a more sophisticated multivariate method may be required. For example, a large Mahalanobis distance from an observation to the mean of all observations is one indication of a multivariate outlier (see Section 5.3), although the truth may just be that the data are not approximately multivariate normally distributed.

It may be difficult to decide what to do about an outlier. Clearly, recording errors or other definite mistakes can be excluded from the analysis. However, if the observation is genuine, then this is not valid. Appropriate action is context dependent. See Barnett and Lewis (1994) for a detailed discussion of possible approaches to the problem.

It is sometimes effective to do an analysis with and without the outliers. If the conclusions are the same, then there is no real problem. If the conclusions depend strongly on the outliers, then they need to be dealt with more carefully.

References

Barnett, V., and Lewis, T. (1994). *Outliers in Statistical Data*. 3rd Edn. New York: Wiley.

Han, J., Pei, J., and Tong, H. (2023). *Data Mining. Concepts and Techniques*. 4th Edn. Cambridge, MA: Morgan Kaufmann.

Hand, D., Mannila, H., and Smyth, P. (2001). *Principles of Data Mining*. Cambridge, MA: MIT Press.

R Core Team. (2024). *R: A Language and Environment for Statistical Computing*. Vienna: R Foundation for Statistical Computing. www.R-project.org/.

Appendix: An Introduction to R

On the website of the Comprehensive R Archive Network (CRAN), www.r-project.org, R (R Core Team, 2024) is described as "a free software environment for statistical computing and graphics . . . maintained by a R Development Core Team." Nowadays, R has taken a leading role in science, business, and technology as a computational system for excellence with data manipulation, calculation of statistical procedures, and graphical displays. You can install R in any operating system (Windows, Mac, or Linux). If you have a Mac or a Windows system, you may want to execute the installer, downloaded from the mirror site of your preference. You just need to follow the directions given in www.r-project.org. Usually, a new version of R is published in March and September each year, but you should be aware of the latest releases announced in the News section of the R-project website.

Here we will provide a brief introduction to R's many aspects:

- Graphical user interface (GUI)
- Help features
- Packages (usually of statistical procedures)
- In more detail, objects (vectors, matrices, and data frames), which are the core of what R works with

A1. The R Graphical User Interface

R has a simple graphical user interface (GUI) that is adequate to input code, to get numerical results displayed in a console, and to generate new pop-up windows (the graphics devices) where graphics are produced. Once R is started, you will see the R Console window containing information about the version of R you are running and other aspects of R. Then, the prompt > appears, indicating that R is in interactive mode. This means that R is waiting for a command to be written and executed. Another way to work in R, the most usual alternative, is through the script editor, accessible from the File menu. Whenever the script editor is invoked, a simple text editor pops up so that you can write sequences of R commands, with the advantage that they can be saved in an ASCII file, readable by any text editor. By using the procedures **Run line or selection** or **Run all**, found in the **Edit** menu, sets of R code or the entire script can be passed to R.

The rudimentary functionality of the standard R GUI makes R unattractive for those users familiar with commercial statistical packages characterized by their friendly interfaces. As improved substitutes for the R GUI environment and its script editor, several applications are offered to facilitate the access to menus and make the task of writing R scripts easier. Possibly in your academic and professional circles a number of these *Integrated Development Environments* (IDEs) for R programming are preferred. R is case-sensitive, and R programming requires the frequent use of parentheses and brackets, and so on. Any IDE for R-coding makes use of colors to indicate corresponding opening and closing parentheses and brackets, as well as to differentiate between sentences from the R language, arguments of functions, and comments. None of these facilities are available in the script editor of the R GUI. You should, therefore, probably choose one of these friendlier applications to improve your interaction with R. Among these IDEs, we strongly recommend the use of Rstudio (Posit Team, 2024).

A2. Help Files in R

R includes a help system, which provides basic information about each R command, such as its syntax, the outcome/object produced, examples of use, and references from where the command is based. It is essential to rely on this standard help system whenever there is doubt about the correct syntax of R statements. In R GUI, you can access the help system as one option of the menu bar, choosing different search strategies. Actually, the help system provides more information than just command syntax. You can also access R documentation and manuals, learn about the development of R, and get answers to frequently asked questions. We can also take advantage of the large number of printed and online resources created by the R community for getting help on a particular command or topic.

A3. R Packages

A package in R is a set of related/integrated R functions, help files, and data sets that either the user or R itself invokes and makes available for a particular purpose. With the exception of a group of packages already installed with R, the user will need to download and install a package of interest into R. Installation is only carried out once, and each time a package of this sort is required, the user only needs to load it into the R environment for the current session. The main public package repository is available in the CRAN mirror sites all over the world, but it is wise to choose the closest location to download and install a package. Accessing the **Packages** menu in R GUI is the best way to select repositories and mirror sites and to run the automatic installation of packages and their so-called dependencies, that is, additional packages necessary for the functionality of the one in which you are interested. By default, a package is stored under a directory called **library** in the R environment, and it contains the necessary procedures, documentation, and connectivity to R. It is also possible to install a package locally (an option present in the **Packages** menu), but its functionality might be affected if any dependency is absent in the set of installed packages. R offers a huge variety of packages. For example, in Version 4.3.3 released in 2024, the CRAN repository contains more than 20,600 available packages. However, in this book, we will just indicate the packages that are necessary to get the results for the examples presented. For R-packages still under development, it is common to host and distribute them in GitHub (https://github.com/), a web-based platform for version control and collaboration. Moreover, some programmers do not intend to submit their packages to CRAN, but only on GitHub. Before installing packages from GitHub, the user needs to install and load the devtools package. See https://cran.r-project.org/web/packages/githubinstall/vignettes/githubinstall.html for more details.

Two other large projects offering package repositories are Bioconductor (Gentleman et al., 2004; https://bioconductor.org/), specialized in providing open-source software for bioinformatics, and R-Forge (https://r-forge.r-project.org/), a central repository for the development of R-packages. The methods of package installation from these two projects are different from the methods described here for CRAN- or GitHub-hosted packages; the exact instructions can be found in the links provided earlier.

A4. Objects in R

The R language manipulates objects. These are defined as entities that can be represented on a computer: numbers, variables, matrices, user functions, data sets (known as data frames in R), or a combination of all these (such as lists). A name can be given to any object, as long as the sequence of characters forming the name does not contain space(s), comma(s), colon(s), semicolon(s) nor any of the following characters:

```
- + * / # % & ( [ { ) ] } ~
```

In addition, object names cannot start with a number, and they are case-sensitive. Thus, X and x are different and refer to different objects. To avoid subsequent typographical errors, it is recommended that you use short and mnemonic object names. Assignment of an object name means that the object is allocated to the current workspace of R via the assignment operator, composed by the character < and a hyphen, with no space between them, i.e., <-. As an example, you may want to define a variable called S carrying the value 12. The command is then S <- 12. If S does not exist, it will be created. Otherwise, its previous contents are replaced. Then, S will be kept in the workspace (in the memory of the computer), but other commands are needed (e.g., save) to store it to a disk. There are also ways to get rid of an object in the current session through the command rm.

A4.1 Vectors in R

The main type of object in R is the vector. It is an n-tuple of objects of the same class. The familiar vector of real numbers is the best example of a vector in R that can be created using the "combine" or "concatenation" function c(). As an example, consider the data shown in Table 1.3. Assume that you would like to analyze the annual precipitation of the sites where colonies of the butterfly *Euphydryas editha* were sampled. To do this, you would ask R to store the values in the numeric vector precip.

```
# Creating the vector precip (annual precipitation on 16 sites)
precip <- c(43, 20, 28, 28, 28, 15, 21, 10, 10, 19, 22, 58, 34, 21, 42, 50)
precip
[1]  43 20 28 28 28 15 21 10 10 19 22 58 34 21 42 50
```

The vector assignment to precip is preceded by a commented line, i.e., a line starting with the hash mark #. Anything written thereafter will be ignored by the R-interpreter. Comments increase the readability of the R-script, both for the programmer and other users. As a vector in R, precip is a single-row array whose length is equal to its number of elements, in this case, 16. You can verify this length by writing

```
length(precip)
[1]  16
```

where the second line gives the response to the first line.

Each element in a vector has its own index or column number, and this can be referred to by enclosing it in square brackets. As an example, the 13th element of precip is precipitation 34 mm. You can display this value in R as

```
precip[13]
[1]  34
```

Names can be given to each element of a numeric vector such as precip. These names can be taken from a character vector (i.e., a vector containing alphanumeric characters enclosed in single or double quotation marks). As

an example, to assign names to the elements of precip using the colony abbreviations as identifiers, the next two commands can be used:

```
COLONY <- c("SS", "SB", "WSB", "JRC", "JRH", "SJ", "CR", "UO",
            "LO", "DP", "PZ", "MC", "IF", "AF", "GH", "GL")
names(precip) <- COLONY
```

precip has now identifiers for each precipitation based on COLONY:

```
precip
SS SB   WSB JRC  JRH SJ  CR  UO  LO  DP  PZ   MC IF AF  GH GL
43 20    28  28   28 15  21  10  10  19  22   58 34 21  42 50
```

The most important feature of vectors is that they can only hold one class
of object (either numbers, strings of characters, levels of one factor, or logical values) but not a mixture of them. In particular, numeric vectors can be
operated on using arithmetic expressions applied element by element. Refer
to Chapter 2 in the manual Introduction to R (Venables et al., 2023) for more
examples and further properties of vectors.

A4.2 Matrices in R

Matrices are described in detail in Chapter 2. These mathematical entities
are composed by a set of numbers arranged in rows and columns to form a
rectangular array. In R, matrices are special arrays of data of the same class,
which is the multidimensional generalization of vectors with multiples. As
an example, assume that a plant ecologist documented the number of mammal species (the species richness) in four sites for two consecutive years and
placed the information in a rectangular array (Table A1):

Table A1. Richness of Mammal Species in Four Sites and Two Consecutive Years

Site	Year 1	Year 2
A	2	2
B	2	1
C	3	5
D	2	1

A matrix associated to these data is as follows:

$$\begin{bmatrix} 2 & 2 \\ 2 & 1 \\ 3 & 5 \\ 2 & 1 \end{bmatrix}$$

There are several ways to build up a matrix like the one presented using R. One possibility is to start with a vector containing all the elements in the matrix and then using the `matrix` function. In order to use this method, we transcribe the eight elements of the matrix column-wise to start with an eight-dimensional vector:

```
RICHNESS <- c(2, 2, 3, 2, 2, 1, 5, 1)
```

The elements of the first column of the matrix determine the first four elements of the vector `RICHNESS`, the elements of the second column compose the next four elements of `RICHNESS`; this process gives also the idea for matrices with a larger number of columns.

Now, we convert this vector to a 4 × 2 matrix, with rows indicating the sites and columns corresponding to years. The matrix `RICHMAT` is filled column-wise, accordingly.

```
RICHMAT <- matrix(RICHNESS, nrow = 4, ncol = 2)
RICHMAT
```

```
      [,1]    [,2]
[1,]    2       2
[2,]    2       1
[3,]    3       5
[4,]    2       1
```

R addresses each element of any matrix by its corresponding row and column position. Thus, the observed richness for Site 3 on Year 2 in `RICHMAT` can be found as

```
RICHMAT[3,2]
[1] 5
```

The elements of a particular row or column can be referred to by omitting the corresponding column and row, after or before the comma, respectively. Thus, the richness for Years 1 and 2 in Site 2 is

```
RICHMAT[ 2,]
[1] 2 1
```

Similarly, the observed richness in Year 1 for all the sites is

```
RICHMAT[, 1]
[1] 2 2 3 2
```

It is also possible to build a matrix with a variation of the method used here, by starting with a vector and then setting the matrix dimension using the dim function; by default, the new matrix will be created column-wise. However, after the application of the dim function, the original vector will be replaced by a matrix with the same name. Therefore, we prefer to name the source vector as RICHMAT:

```
RICHMAT <- c(2, 2, 3, 2, 2, 1, 5, 1)
dim(RICHMAT) <- c(4, 2)
```

The right-hand expression defining the dimension of the matrix is a two-element vector c(4,2), specifying the number of rows (4) and columns (2) of the new matrix RICHMAT.

```
RICHMAT
       [,1]     [,2]
[1,]    2        2
[2,]    2        1
[3,]    3        5
[4,]    2        1
```

In R we can implement convenient data-annotation practices so that the data stored and analyzed look like data we recorded in detail. In the example, column names in matrix RICHMAT can be defined or later invoked by means of the command colnames. Similarly, the row names in RICHMAT can be created or displayed using the command rownames. Initially, colnames and rownames indicate that RICHMAT does not have explicit row and column names:

```
colnames(RICHMAT)
NULL
rownames(RICHMAT)
NULL
```

We now create the labels for RICHMAT, as indicated by Table A1, using character vectors:

```
colnames(RICHMAT)<- c("Year 1", "Year 2")
rownames(RICHMAT) <- c("Site A", "Site B", "Site C", "Site D")
```

When RICHMAT is invoked again, we get a better-annotated matrix:

```
RICHMAT
              Year 1      Year 2
Site A          2           2
Site B          2           1
Site C          3           5
Site D          2           1
```

A third method to build a matrix is by combining two or more vectors of the same length (and class). Following a similar idea from the previous example, assume that there are two vectors of four entries each, with the first vector corresponding to the richness found in the four sites for Year 1 and the second one containing the corresponding richness for Year 2, defined as follows:

```
Year_1 <- c(2, 2, 3, 2)
Year_2 <- c(2, 1, 5, 1)
```

The vectors can then be combined (bound) in one matrix using cbind() (combine by columns) or rbind() (combine by rows). In our example, combining by columns is the convenient function to apply:

```
RICHMAT1 <- cbind(Year_1, Year_2)
```

so that

```
RICHMAT1
       Year_1      Year_2
[1,]       2           2
[2,]       2           1
[3,]       3           5
[4,]       2           1
```

In this case, R assigns column names to the matrix RICHMAT1 taken from the source vectors Year_1, Year_2. The following code produces the same output (namely, the first column of RICHMAT1):

```
RICHMAT1[, 1]
[1] 2 2 3 2

RICHMAT1[, "Year_1"]
[1] 2 2 3 2
```

When RICHMAT1 was created, it did not have row names. This can be checked:

```
rownames(RICHMAT1)
NULL
```

Suitable row names can then be given to RICHMAT1:

```
rownames(RICHMAT1) <- c("Site A", "Site B", "Site C",
   "Site D")
```

Additional properties of arrays and matrices can be found in Venables et al. (2023, chapter 5) and in Chapters 2 and 3.

A4.3 Data Frames

To manipulate multivariate data and handle different sorts of variables, R has a special class of object called a *data frame*. Data frames and two-dimensional arrays (matrices) have one property in common: the data are arranged in rows and columns. However, the rows of a data frame are identified as different sampling units from where observations or measurements are made, with these being placed in different columns, so that each column in a data frame corresponds to a particular variable. Therefore, unlike arrays that allow one type of object only, data frames may include a mixture of variables: numeric, character, factor, logical, dates, and so on.

One explicit way of constructing a data frame is to combine different vectors, presumably of the same length and each one containing a particular variable, and declare them in the data.frame command:

```
my.data <- data.frame(vector1, vector2, ...)
```

A better practical way of making a data frame takes advantage of R being capable of reading data files in different formats, such as tab- or comma-separated values (csv), which is the most convenient format for importing data into R. Alternatively, data saved in a spreadsheet and statistical applications such as Excel, SAS, or Minitab can be imported, in which case it is necessary to install suitable R packages to read the data. Here, we will only consider the making of a data frame using the command read.table applied to an external file saved with tab-separated values or csv. As an example, consider the data described in Example 1.5. These data can be accommodated in a data frame with 38 rows (European countries) and seven variables: a non-numeric variable indicating a subdivision of countries with respect to their membership of the European Union, and the cost of six individual food groups that make up the healthy diet basket.

Assume that the data on cost and affordability have been saved as a tab-delimited file with the name **Cost_food.txt**, where this file can be found at the website at the book's website (resource materials section). The first two and last two lines and headings of each column of the data are then

Country	Group	CSTS	CASF	CLNS	CVEG	CFRT	COFT
Albania	NEU	0.599	1.204	0.441	0.707	0.911	0.089
Austria	EU95	0.229	0.612	0.360	0.698	0.819	0.055
⋮	⋮	⋮	⋮	⋮	⋮	⋮	⋮
Switzerland	NEU	0.212	0.675	0.249	0.743	0.577	0.067
United Kingdom	EU95	0.147	0.494	0.195	0.399	0.533	0.053

Here, Country is not a variable; it refers to row names. This can be taken into account when importing the data using read.table. The second column is the qualitative variable `Group`, and the last seven columns are all numeric variables. The command needed to produce a data frame from the tab-delimited file `Cost_food.txt` is then

```
Euro.food <- read.table("Cost_food.txt", header = TRUE, row.names
                  = 1, sep = "\t", stringsAsFactors = TRUE)
```

The execution of `read.table` assumes that a working directory has been chosen in R such that the directory (folder) contains the file **Cost_food.txt**. One way to accomplish this is by accessing the menu **File > Change dir ...** in R GUI. The first argument is the tab-delimited file name enclosed in quotation marks. After the file name, the option `header = TRUE` was written, which means that the file contains a header (names assigned to each column). However, the first column is not a variable; it refers to row (sampling unit) names. This explains the option `row.names = 1`, meaning *the row names are located in column 1*. The argument `sep = \t`, designates a tab as the delimiter of each field in the data set. Finally, the argument `stringsAsFactors = TRUE` in `read.table` converts character vectors (alphanumeric variables) to *factors*. In the earlier example, the vector `Group` is originally a character vector and remains like that in the source file **Cost_food.txt**, but in the importing process leading to the creation of the data frame `Euro.food`, that variable has been changed to factor. A factor is an object (a vector) useful to group the components of other vectors of the same length. The main difference between a factor vector and a character vector is that the components of the former are not enclosed in quotation marks. Internally, R handles a factor as a numeric (integer) vector, i.e., the *mode* of a factor is numeric. The default order of factor levels is dictated by the alphabetic order of the labels. In terms of computer memory usage, storage of factors is better, because an integer uses fewer bytes of memory than a string of alphanumeric characters.

Once the data frame `Euro.food` is created, it contains the same information as the original source file, but it is now accessible to R. To confirm `Euro.food` is indeed a data frame, you can type

```
class(Euro.food)
[1] "data.frame"
```

You can then display the whole data frame in the console by typing

```
Euro.food
```

or the first six rows

```
head(Euro.food)
```

Six rows are the default number of data frame lines shown by the `head` function. There is also a `tail` command that can be used to display the last rows in a data frame (again, six is the default). In fact, `head` and `tail` are not only applicable to data frames but also to other objects, e.g., vectors and matrices.

One helpful way to recognize the type of variables present in a data frame is through the `str` command (i.e., the data frame str*ucture*). In this case,

```
str(Euro.food)
'data.frame': 38 obs. of 7 variables:
$ Group : Factor w/ 4 levels "EEA","EU17","EU95",..: 4 3 4 3 4 2 2 2 3 2 ...
$ CSTS  : num 0.599  0.229   0.506   0.222  0.54   0.504 0.496 0.297 0.235 0.333 ...
$ CASF  : num 1.204 0.612 0.826 0.745 0.921 ...
$ CLNS  : num 0.441   0.36    0.643   0.369   0.512   0.478 0.556 0.377 0.263 0.671 ...
$ CVEG  : num 0.707 0.698 0.349 0.677 0.825 ...
$ CFRT  : num 0.911   0.819   0.678   0.79    0.916   0.959 0.95  0.693 0.546 0.742 ...
$ COFT  : num 0.089   0.055   0.174   0.06    0.132   0.101 0.073 0.076 0.053 0.085 ...
```

The imported variables are listed in different lines, each starting with a dollar sign. This pinpoints the way each variable contained in a data frame is referred to in R. The name of the variable is preceded by the name of the data frame, and these two are separated by the $ character. Thus, to display the variable (vector) `CFRT` (the purchasing power paid per person per day to buy fruits in the European countries) present in `Euro.food`, the command is

```
Euro.food$CFRT
 [1] 0.911 0.819 0.678 0.790 0.916 0.959 0.950 0.693 0.546 0.742 0.597 0.859
[13] 0.714 0.581 0.828 0.484 0.621 0.812 0.648 0.670 0.762 0.931 0.833 0.607
[25] 0.640 0.615 0.651 0.687 0.432 0.697 0.745 0.872 0.634 0.669 0.796 0.773
[37] 0.577 0.533
```

An error message is produced if you request a variable without a data frame name:

```
CFRT
Error in eval(expr, envir, enclos): object 'CFRT' not found
```

Further properties of variables within a data frame can be seen by specific R commands. Thus, the vector `Euro.food$Group` is of *factor class*, as shown by

```
class(Euro.food$Group)
[1] "factor"
```

but of *numeric mode*:

```
mode(Euro.food$Group)
[1] "numeric"
```

The strict rule of writing the data frame name and the variable name with a dollar sign in-between can be avoided with suitable code executed before calling the variable, such as using the commands `attach` or `with` (see

Venables et al, 2023, chapter 6). It is important to bear in mind that a data frame is just a special case of a more general data structure in R known as *list*, which is considered one of the most versatile objects in R for handling data or the results produced by other R functions. See Davies (2016, chapter 5) for more information about lists.

Regarding the example of cost and affordability of healthy food in European countries, the data could also be stored as a csv file. Assuming these data originally saved as **Cost_food.csv**, the R command that can be used for making a data frame (named `Euro.food1`) would then be

```
Euro.food1 <- read.table("Cost_food.csv", header = TRUE,
                    row.names = 1, sep = ",", stringsAs
                    Factors = TRUE)
```

The separator of each field is now a comma (`sep = ","`). Alternatively, a csv file can be read instead by the command `read.csv`. It is similar to `read.table`, except that a comma is the default separator for the file to be imported; therefore the `sep = ","` argument can be omitted:

```
Euro.foodcsv <- read.csv("Cost_food.csv", header = TRUE,
                    row.names = 1, stringsAsFactors =
                    TRUE)
```

See chapter 7 in Venables et al. (2024) for further topics related to reading data from files.

A4.4 Data Frame Indexing (Single, Double, and Logical)

Data frames share the properties of single and double indexing described for vectors and matrices. Thus, specific values in an individual vector (variable) within a data frame can be accessed by referring to only one index (indicating the position of the datum or the data for that variable), whereas the access of groups of variables and/or groups of rows is possible through double indexing (of rows and columns) or logical operators. As a first example, the contents of the variable `CFRT` in the data frame `Euro.food` for countries occupying rows 3 to 8 can be verified as

```
Euro.food$CFRT[3:8]
[1]  0.678 0.790 0.916 0.959 0.950 0.693
```

where `3:8` is R shorthand for the vector `c(3, 4, 5, 6, 7, 8)`. Similarly, the cost of fruits for people living in the first, third and fifth countries as they appear in the data frame are

```
Euro.food$CFRT[c(1, 3, 5)]
[1]  0.911 0.678 0.916
```

For each of the two last examples, a single vector within the data frame `Euro.food` was called, and the corresponding output was also a vector. In other situations, like those about double indexing shown in the following, the selection of larger data frame portions (more than one single vector) results in a subset of the original data that is also a data frame. In case of doubt, verify the class of an object with the `class` command. Should you need to *coerce* (convert) a vector to a data frame, the explicit *"as-dot"* function `as.data.frame()` is useful. The next three sentences in R coerce into a data frame the vector produced in the last example, and give explicit row and column names for the "sub-data frame" CFRT3C:

```
CFRT3C <- as.data.frame(Euro.food$CFRT[c(1, 3, 5)],
              row.names = rownames(Euro.food)[c(1, 3, 5)])
names(CFRT3C) <- "CFRT"
CFRT3C
                       CFRT
Albania                0.911
Belarus                0.678
Bosnia and Herzegovina 0.916
```

See Davies (2016, chapter 6) and the R-help system for more details about "as-dot" coercion functions.

The following three examples of double-indexing produce data frames as objects

```
Euro.food[25:30,]
                    Group CSTS  CASF  CLNS  CVEG  CFRT  COFT
North Macedonia     NEU   0.525 0.949 0.347 0.712 0.640 0.146
Norway              EEA   0.565 0.691 0.257 1.122 0.615 0.075
Poland              EU17  0.318 0.584 0.377 0.890 0.651 0.090
Portugal            EU95  0.271 0.583 0.275 0.620 0.687 0.078
Republic of Moldova NEU   0.462 0.746 0.360 0.339 0.432 0.121
Romania             EU17  0.358 0.945 0.351 0.497 0.697 0.073
```

```
Euro.food[c("United Kingdom", "Romania"),]

               Group CSTS  CASF  CLNS  CVEG  CFRT  COFT
United Kingdom EU95  0.147 0.494 0.195 0.399 0.533 0.053
Romania        EU17  0.358 0.945 0.351 0.497 0.697 0.073
```

and

```
Euro.food[, c(2,5)]

               CSTS  CVEG
Albania        0.599 0.707
 ⋮
United Kingdom 0.147 0.399
```

which is asking for the second and the fifth variables (for the sake of space only the first and the last rows are shown). Also,

```
Euro.food[, c("Group", "CFRT")]
                        Group   CFRT
Albania                 NEU     0.911
:
United Kingdom          EU95    0.533
```

produces a "sub-data frame" containing variables Group (the EU or non-EU group) and CFRT (the cost of fruits), with only the first and last lines shown here.

As a final example, the following command involves a logical operation:

```
Euro.food[Euro.food$Group == "NEU",]
                        Group CSTS  CASF  CLNS  CVEG  CFRT  COFT
Albania                 NEU   0.599 1.204 0.441 0.707 0.911 0.089
Belarus                 NEU   0.506 0.826 0.643 0.349 0.678 0.174
Bosnia                  NEU   0.540 0.921 0.512 0.825 0.916 0.132
 and Herzegovina
Montenegro              NEU   0.425 0.862 0.367 0.829 0.833 0.081
North Macedonia         NEU   0.525 0.949 0.347 0.712 0.640 0.146
Republic                NEU   0.462 0.746 0.360 0.339 0.432 0.121
 of Moldova
Russian Federation      NEU   0.360 0.980 0.191 0.739 0.745 0.135
Serbia                  NEU   0.572 0.998 0.612 0.920 0.872 0.096
Switzerland             NEU   0.212 0.675 0.249 0.743 0.577 0.067
```

The output displays a data frame for those sampling units (rows) in the data frame Euro.food fulfilling the (logical) condition that the variable Group is equal to "NEU". The double equals sign == is the way R specifies a conditional or logical equality sign, so that the rows shown are only those for which it is TRUE that the Group is equal to "NEU".

For further examples of the use of single/double indexing or logical expressions, as ways to extract portions of data frames, see Davies (2016, chapter 5).

A5. Plots in R

Functions for base *high-level plotting* (Venables et al., 2023, chapter 12) are already installed in R in the graphics package. The type of plot these functions produce is dependent on the class of the objects written as argument(s) (Venables et al., 2023, chapter 12). For simple scatterplots, the default scatterplot function called plot.default can be invoked with the plot command.

```
plot(x, y, type = "p", ...)
```

Here x and y specify coordinates for the points to plot; usually they are simply numeric data vectors. The type = "p" argument is the default specification if a scatterplot is desired (it can be omitted in this case), indicating that points will be shown, y on the vertical axis and x on the horizontal axis. Other parameters controlling decorations of the plot such as axes, titles, point shapes, point sizes can be written as additional arguments of plot. The advantage of the high-level plotting functions like plot is that default decorations are automatically generated, producing suitable basic plots that can be later improved with the addition of more arguments to plot and the inclusion of *low-level plotting* commands.

A more convenient way to invoke the function plot, especially when the data are numeric vectors inside a data frame, involves the operator ~, used in other functions to define a model formula in R. This specification is known as a *formula notation* for scatterplots and its syntax takes the form:

```
plot(formula, data = df, ...)
```

The argument formula is an expression such as y ~ x1 + x2 + ... , usually interpreted as a response variable y being modeled by linear predictors. Thus, the notation y ~ x indicates that x is as a single predictor of y. However, the scope of this formula transcends to cases in which there is no interest to model y as a function of x but simply seeks to draw a scatterplot, putting y on the vertical axis and x on the horizontal axis. The object df in the presented formula specifies a data frame (or list) from which the variables in formula should be taken.

At startup, R starts a graphics device driver which opens a special graphics window every time a high-level plotting function is called. Although this is done automatically, a graphics window is opened when X11() under UNIX, windows() under Windows or quartz() under macOS is executed. This window (whose nominal size is 7 × 7 inches) has a menu of options, including facilities to save the plot directly from the clipboard or to a file. Here we illustrate a simple way to produce in R the plot shown in Figure 3.1, by passing suitable arguments to the plot function.

First, the sparrow data is imported and made available in the R workspace:

```
sparrows <- read.csv("Bumpus_sparrows.csv", header = TRUE,
                     stringsAsFactors = TRUE)
```

A simple scatterplot of alar extent against the total length for the 49 female sparrows can be generated with the plot command.

```
plot(sparrows$Total_length, sparrows$Alar_extent)
```

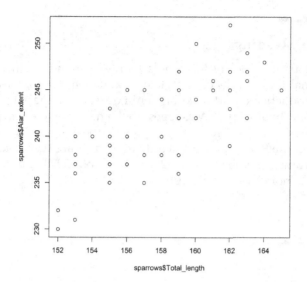

FIGURE A.1
A simple scatterplot of alar extent against the total length for the 49 female sparrows, using two vectors as arguments in the `plot` command. See Figure 3.1.

Alternatively, we can use the "formula" version of the `plot` function (try it!).

```
plot(Alar_extent ~ Total_length, data = sparrows)
```

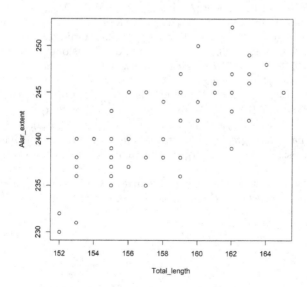

FIGURE A.2
A simple scatterplot of alar extent against the total length for the 49 female sparrows, using a model formula as argument of the `plot` command. See Figure 3.1.

A6. Further Reading

The ubiquitous presence of R, both as a programming language and as a computational tool in science, technology, and humanities, have led to an explosive production of textbooks and online resources specialized in teaching beginners about its use. We suggest that a narrow web search, suited to the reader's needs and interests, would provide the best way to find good references about the R system. Our favorite references are the books by Davies (2016), Ekstrøm (2016), Jones et al. (2022), Kabacoff (2020), Long and Teetor (2019), and Zamora Saiz et al. (2020).

References

Davies, T.M. (2016). *The Book of R. A First Course in Programming and Statistics*. San Francisco, CA: No Starch Press.

Ekstrøm, C.T. (2016). *The R Primer*. 2nd Edn. Boca Raton, FL: Chapman and Hall.

Gentleman, R.C., Carey, V.C., Bates, D.M., Bolstad, B., Dettling, M., Dudoit, S., Ellis, B., Gautier, L., Ge, Y., Gentry, J., Hornik, K., Hothorn, T., Huber, W., Iacus, S., Irizarry, R., Leisch, F., Li, C., Maechler, M., Rossini, A.J., Sawitzki, G., Smith, C., Smyth, G., Tierney, L., Yang, J.Y.H., and Zhang, J. (2004). Bioconductor: open software development for computational biology and bioinformatics. *Genome Biology* 5: R80. https://doi.org/10.1186/gb-2004-5-10-r80.

Jones, E., Harden, S., and Crawley, M. (2022). *The R Book*. 3rd Edn. Chichester: Wiley.

Kabacoff, R.I. (2020). *R in Action*. 3rd Edn. New York: Manning.

Long, J.D., and Teetor, P. (2019). *R Cookbook*. 2nd Edn. Sebastopol, CA: O'Reilly.

Posit Team. (2024). *RStudio: Integrated Development Environment for R*. Boston, MA: Posit Software, PBC. www.posit.co/.

R Core Team. (2024). *R: A Language and Environment for Statistical Computing*. Vienna: R Foundation for Statistical Computing. www.R-project.org/.

Venables, W.N., Smith, D.M., and the R Core Team. (2024, April 24). *An Introduction to R: Notes on R, A Programming Environment for Data Analysis and Graphics, Version 4.4.0*. https://cran.r-project.org/doc/manuals/R-intro.pdf (Accessed: 15 June 2024).

Zamora Saiz, A., Quesada González, C., Hurtado Gil, L., and Mondéjar Ruiz, D. (2020). *An Introduction to Data Analysis in R*. Cham: Springer.

Index

Note: Page numbers in *italics* indicate a figure and page numbers in **bold** indicate a table on the corresponding page.

Index of R Functions and Packages

Note: Page numbers in *italics* indicate a figure and page numbers in **bold** indicate a table on the corresponding page.

Printed in the United States
by Baker & Taylor Publisher Services